T0064180

Loving Math: The Sequel

Ever More Challenging Problems, Solutions and Discussions With Special Applications

by

Lewis Forsheit

Order this book online at www.trafford.com
or email orders@trafford.com

Most Trafford titles are also available at major online book retailers.

Print information available on the last page.

ISBN: 978-1-4907-7915-7 (sc)
ISBN: 978-1-4907-7916-4 (e)

Library of Congress Control Number: 2016920257

Trafford rev. 12/08/2016

www.trafford.com
North America & international
toll-free: 1 888 232 4444 (USA & Canada)
fax: 812 355 4082

Table of Contents

Contents

Introduction

In 2004, I wrote and published Loving Math, *Advanced Problems with Solutions, Applications and Comments*. I did so because I had recently begun my teaching career and was experiencing not only the passion of the mathematics that I loved but also the students' excitement and infectious spirit of youth. I wanted to share the knowledge and joy of that experience.

After nine years, when I retired, I can truly say that my feelings were even more positive and that I had learned a lot. I still continue with my individual math mentoring and I keep on exploring new ideas. I also had the benefit of working in a school whose pedagogy was project-based learning. This gave me the opportunity to have the students work on a wide range of problems and projects beyond the regular syllabus.

This book includes problems and projects (either in whole or in part) that my students worked on in my regular math classes, in my advanced math elective, or outside of school with my mentoring. The problems and projects include some that can be found in the literature, some that I made up myself, and some (no doubt) that are both because what I made up may have been made up previously.

There are references to the *Loving Math (LM)* book throughout this one. If you have *LM*, you can see the connections; if not, you can still proceed. In any event, if there is anything incorrect herein, it is on me. It is bad enough to make errors, but putting them in print only makes them worse. Speaking of which, I can report two typos in the *LM* book. First, the word "cerians" should be "cevians" on pages 4, 5, 29 and 88. Second, the equation on page 100 should not include
$4r^2 - a$ and $4r^2 - b$, but rather $4r^2 - a^2$ and $4r^2 - b^2$.

I decided to compile some of the topics that my best students worked on and publish them for the benefit of other teachers, who could share them with their gifted students. Some of the problems and projects are presented with complete solutions with every step detailed. Some are missing a number of the intermediary steps, which need to be filled in, and some are problems and projects that are only proposed without any solution offered.

Specifically, here is a rundown on the problems and projects I have included.

1. Equal Angle Bisectors Theorem Revisited

   This theorem was presented in *LM* p 96 with its classical proof.

   I am offering a different proof here.
2. Right Triangle with Inscribed Circle Quadrant

   The solution set up here is easy, and we find a closed-form solution to a quartic equation.
3. Ten Proofs of the Cosine of an Angle Sum

   My pedagogy has always highly valued multiple perspectives.

   Students should enjoy the pursuit of as many proofs as possible.
4. Simple Estimate for a Cubic Equation Root Using the Golden Ratio

   We always studied the Golden Ratio; here's another approach.

   The results here may be original; and make cubics easy to estimate.
5. Maneuvering Around a Corner

   This calculus extrema problem has a great answer that I love.

   I made up the problem, but I can't image it hasn't been done before.
6. Calculus of Comparing Investment Performances

   This problem can be presented as a pure calculus extrema problem.

   But, I like putting it into a real-world application context.
7. Calculating Important Parameters of an Investment (Retirement) Account

   What can I say? It came to me while working with my financial adviser. Perhaps the financial industry can use it to develop apps.
8. Using Origami's Third Dimension to Double a Volume

   This is a classical proof of the geometry of a folded plane.

   The idea of using another dimension to solve an old problem is key.
9. Solving for Equally Damped Sinusoids

   It's a differential equation opening, but it is solved with algebra.

   I published this on an online physics-forum website.
10. Raising a Complex Number to a Complex Number

    This is a classical algebra problem. I worked with a student using this approach on Riemann's Hypothesis, which we've published.
11. Human Modeling Using Imaginary Numbers

    All my students, and probably my readers, too, know that I love imaginary numbers. Here's a novel application on the human side.

In addition to these problems, which are solved, there are several other problems and projects that are not solved but are suggested for future work. These include: a Golden Ratio approach to a variant of the equation studied here, a Golden Ratio approach to studying trapezoids, passing through hallways with rounded conical corners, using Origami to address trisecting an angle and squaring a circle, Mandelbrot fractals, the Schrodinger Equation, and the Riemann Hypothesis. Over the years, my students have addressed these topics in their math-class projects.

I have also included a Forward, which is a short history of real and imaginary numbers, with an eye toward how imaginary numbers can be used in the future. This tableau is designed to give both experienced and lay persons some background about how we have used numbers and how we may use them.

The last problem, Section III Problem 11, is of particular interest in the pursuit of the potential that imaginary numbers hold. Are they "real?" Do they exist? Where can I see them in the world? I believe the pursuit of these studies will yield great rewards.

I would like to thank all the students I have had, my colleagues, and Wildwood School, where I taught. Also, I would like to give a special shout-out to my mother, Nettie Forsheit, who has always been very supportive of all my efforts and is now celebrating her 98th birthday — quite a mathematical feat. Special thanks go to Karre Jacobs, who edited the narrative and, as always, my wife, Arleen, continues to be my inspiration and has helped throughout the book's development. We have been enjoying our retirement and, at the time of this printing, our 52nd year of marriage.

Good luck and enjoy.

Lewis Forsheit
Los Angeles
lew_forsheit@earthlink.net
January 2017

Forward

Real and Imaginary Numbers

The purpose of this historical review is to prepare the reader (especially a novice) for the progression from the early problems that use real numbers through the later problems that use imaginary numbers and, finally, to the last problem, which applies imaginary numbers to a human application in a novel way.

Appreciating Real Numbers

Since the dawn of human existence, we have been counting. Mostly, we have been counting what we saw and what we had; it was a survival need. As a basic element of life, the counting of numbers became real, though they were, really, just a concept. There is no such thing as "a three." There are, however, three things that can be counted.

The ancients represented these counting numbers with pictures or even alphabet letters. But it did not facilitate the daily-calculation needs of an emerging species; even the Romans could not calculate well with Roman numerals. The Arabs (and Indians) are credited with developing a way of depicting numbers that did facilitate calculations. Moreover, they are credited with the invention of the number "zero" as a "place holder." So, we call these tools "Arabic numerals."

We also became aware of fractions, or what we now call rational numbers. Most people got the hang of this quickly. Perhaps there was only half a carcass to eat; perhaps there were only a tenth of food reserves remaining. But there were conceptual issues to confront. When you cut an apple in two to share, each has half an apple. But when you break a rock in two, each still has a rock (though smaller).

Whereas rational numbers became more intuitively obvious over time, irrational numbers, which are those that cannot be calculated as a fraction, were a challenge. The ancients became aware of this and they coped well. When a tract of land to be claimed needed to be twice the area of a previous one, they did this by constructing an isosceles right triangle and used the hypotenuse, rather than a leg, and the job was done. The Greeks actually could prove that these numbers, which became important in construction, were not rational fractions. (*LM* p 118) So, there was an element of disbelief that they could really be numbers; but, on the other hand, they could be drawn and so they had their place.

Initially, irrational numbers lived only in the domain of geometry, but with the ascendency of algebra, they were found to be the solutions to polynomial equations. Transcendental numbers then emerged as those that could not; the most famous one of these was pi. Although pi was measurable as the ratio of a circle's circumference to diameter, the measurement could not be done accurately, and it would take a long time to figure out what it was all about. But, still, we could look at a wheel and understand pi.

One of the great challenges humankind faced was negative numbers. We were familiar with subtraction since we somehow needed to express the idea that we had less now than we had before. But, could we lose more than we had? Two minus one made sense; one minus two made no sense. Negative numbers were reviled by some and feared by others.

In time, though, negative numbers were accepted because they could represent a debit or a debt, something that was owed. And with the use of coordinate axes, negative numbers could model the differences between left and right, up and down, front and back.

At some point, a creative new idea emerged, called the "number line." This was a graphical, pictorial representation of all the numbers. There was a place for every number; and every number was in its place. Every number was understood, and, each in its own way had a purpose and was useful. Later these number lines could be used as axes, à la Descartes, so that we could locate ourselves and solve problems that involved the movement of multiple objects.

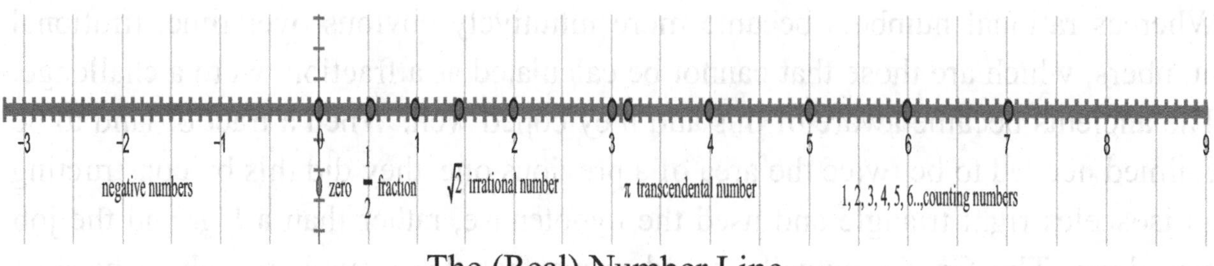

The (Real) Number Line

Appreciating Imaginary Numbers

Even in ancient times, equations were being created to represent the physical world people found themselves in. The Golden Ratio, as a solution to a quadratic equation, is a famous example that Greeks loved due to its aesthetic. Quadratic equations were those that had an x^2 term as its highest order term; that is, no x^3, x^4, etc., terms.

And, sometime in the first millennium, the quadratic equation was formally solved. Namely, for the equation

$$ax^2 + bx + c = 0,$$

the two solutions were

$$x = \frac{-b \pm \sqrt{b^2 - 4ac}}{2a}.$$

Usually, only the positive root was usable as valid, but with the increasing awareness of negative numbers, the negative solution could also be used, in proper context. The term "extraneous" describes a solution that, although it satisfies the equation, is not appropriate to the problem's context. An example would be getting a negative solution for the length of a plot of land.

But, then it happened. And it happened repeatedly Some equations' solutions had a negative number under the square root sign. In fact, these equations had no other solutions (also called "roots") other than these. Everyone ignored these solutions because they made no sense. There were no numbers, even negative numbers, that when multiplied by themselves resulted in a negative number. And, there was no place for them on the number line. After hundreds of years, a compromise was reached: The square root of a negative number would be called "imaginary," meaning that no one believed there was any reality to them, in exchange for continuing to use them to solve equations that could not be solved any other way. An imaginary number such as $\sqrt{-4}$ was now written $\sqrt{4}\sqrt{-1} = \pm 2i$ and the new unit of imaginary numbers was called "i." And "complex" numbers were a combination of a "real" number and an "imaginary" number such as $3+4i$.

Around the turn of the twentieth century, two important developments occurred. One was that the problem of locating imaginary numbers on the number line was solved. The other was that several important, real-world applications started using imaginary numbers.

To address the location question, it was noticed that if a positive number is multiplied by -1, then its position rotated 180 degrees around the zero point on the number line. So if i multiplied by i is -1, then a multiplication of a real number by i would result in a 90-degree rotation around the zero point.

With this advent, the "real number line," as we knew it, now became the "complex number plane." (See figure below.)

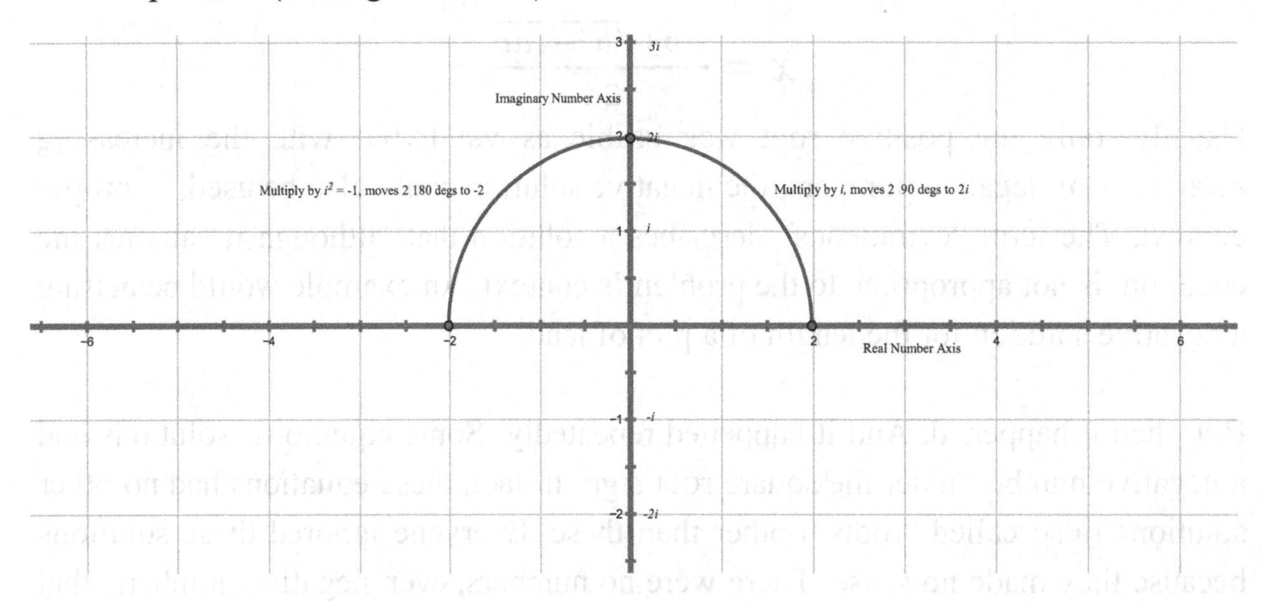

Because the imaginary axis is perpendicular to the real axis, it became a good tool for modeling natural phenomena that had perpendicular characteristics. One is electromagnetic waves, where the electric field is perpendicular to the magnetic field and both are perpendicular to the direction they are traveling. Imaginary numbers helped model and solve the macro phenomena, as well as at the quantum-mechanics level. Another is electric AC circuits (*LM* p 124), where voltage and current are often out of phase by some number of degrees (though not usually ninety).

Another key development was Einstein's relativity theory. The General Theory of Relativity uses the concept of real time as the fourth dimension. Subsequently, the Special Theory of Relativity uses the concept of imaginary time as the fifth dimension. These formulations helped explain some of the great mysteries of the universe.

Imaginary time is perpendicular to real time: Whereas real time is sequential, meaning that it only increases; imaginary time is concurrent, meaning it can go forward or backward. Therefore, at any point in real time, imaginary time has access to all past, present and future time. Later, Stephen Hawking popularized the concept of imaginary time by proposing that by using imaginary time, the big bang creation of the universe did not occur.

Over the years, I tried to convince my students that having a good knowledge of imaginary numbers could get them a good paying (engineering) job and that that would truly be "real." But though imaginary numbers could solve equations, model real-world phenomena and lead to good jobs, the basic problem remains: You can't count any real-life objects using i. Thus, imaginary numbers remain only a tool or a mathematical trick.

Just as irrational, transcendental and negative numbers took many years to be accepted, so too must imaginary numbers go through this process. But the challenge is greater, since it is so difficult to experience them in our tangible lives. It is left for gifted students to take on this challenge and bring significant meaning to this next numerical frontier. The following sections make suggestions as to how this can be done.

In this book, Section II has solutions to the first 10 Problems. The first eight require only real numbers. Problems 9 and 10, as well as their three associated unsolved problems, require the use of imaginary numbers for solving. These problems also encompass many of the practical applications of imaginary numbers currently in use.

The last problem, Problem 11, is in Section III, alone, so we can focus on how imaginary numbers can be extended to the frontier of human self-exploration.

The Outlook for Imaginary Numbers

Our search for imaginary numbers may have been thrown off course due to the name "imaginary," which suggests they may not exist. It may be more "realistic" to call them "intangible" numbers.

Hopefully, we can agree that intangibles exist and are real. Indeed, it is said, that the best things in life are not things. In this book, it is posited that there are three categories of intangibles: values, ideas and emotions; which correspond with what we accept without proof, what we analyze with our left brain, and what we understand holistically with our right brain, respectively.

Values include: existence itself, articles of religious faith, political and societal norms, the scientific method, etc. Ideas include: laws of mathematics and science that we can prove (based on our assumptions/values), innovations in technology, managing money and resources, planning projects, etc. Emotion is probably the widest-ranging intangible, including: love, courage, patriotism, ambition, hope, caring, creativity, anger, fear and many more. Most people express that it is this last category that is most important in their own lives.

Perhaps we can use imaginary (intangible) numbers to measure the intangibles with which we are already familiar. As it is, we have no other way to measure them. "How do I love thee, let me count the ways" is poetic (right brain) but hardly comes close (left brain). A business manager once defined "a good idea is one that rolls out the factory door," which is, at least, measurable but is a rough-cut approximation, at best.

One possible model is to consider the three imaginary axes, x_i, y_i and z_i as a way to locate and measure our values, ideas and emotions, respectively. Adding these new dimensions is comparable to the addition of imaginary time to help explain relativity.

Sometimes when we are asked "where are you?" we can give them a location, which can be drawn on a Cartesian grid. But, many times, the question is asking us about ideas we are pondering, plans we are making, and emotions we are experiencing. Are these not just as real?

Perhaps with eight axes, we can describe our own life existence and those of others. We could understand better "what makes us tick" and how our thoughts and actions are more likely synchronized.

Consider the track record of imaginary or intangible numbers. They began as a concept needed to solve equations without any conceivable meaning or use in the "real" world. In time, they added a new dimension to the way we understand numbers just as Origami added long-sought, two-dimensional solutions by adding a third dimension through folding. Imaginary numbers' inherent property of perpendicularity describes how electric circuits work and how the inner secrets of quantum mechanics operate. Imaginary numbers are the pathway to, arguably, the most aesthetic infinite fractal graphic design, as well as the world's most perplexing math problem. It should not be surprising that imaginary numbers could become the basis of humankind's next great leap in our consciousness.

Perhaps, the reason we have not found imaginary numbers in the real world is that we were looking outward rather than inward. The answer may have been with us all the while and we did not see it. Section III addresses how such imaginary number coordinate axes might work.

Gifted students, in the early–twenty-first century, have the opportunity to explore this new-number frontier and, hopefully, open up new vistas for humankind, as did each of the then new numbers of the past.

Section I

Problem Statements

1 – 11

1 Equal Angle Bisectors Theorem Revisited

Prove that if two angle bisectors of a triangle are equal, then the triangle is isosceles.

See *LM* p 96 for the classical Steiner-Lehmus proof.

The objective now is to find another proof.

Hint: Use Stewart's Formula for angle bisectors. *LM* p 90.

2 Right Triangle with Inscribed Quarter Circle

Find the length of the radius, r, of a quarter circle such that when given line segments a and b are added to r to form the legs of a right triangle, the quarter circle is also tangent to the hypotenuse.

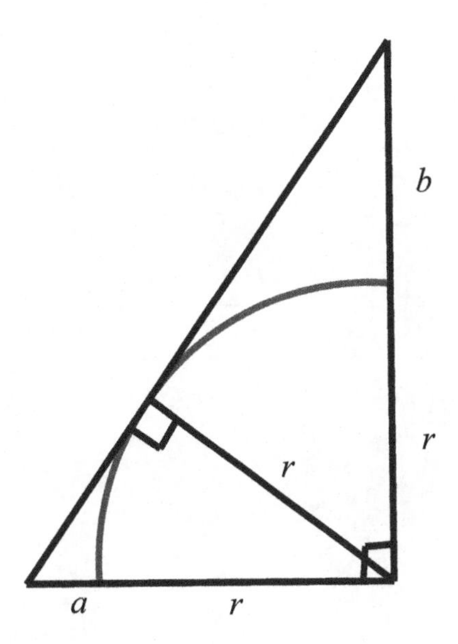

3 Ten Proofs of the Cosine of an Angle Sum

Present 10 proofs of the trigonometric identity:

$$\cos(\alpha + \beta) = \cos\alpha \cos\beta - \sin\alpha \sin\beta$$

4 Simple Estimate for a Cubic Equation Root Using the Golden Ratio

Given the cubic equation

$$y^3 + py^2 + qy + r = 0$$

We know that if we let $y = x - \frac{p}{3}$
then the "depressed" or "reduced" equation is

$$x^3 - 3cx - 2a = 0$$

where $c = \frac{p^2}{q} - \frac{q}{3}$, $a = \frac{pq}{6} - \frac{p^3}{27} - \frac{r}{2}$ and $b = a^2 - c^3$.

A real root of this equation is

$$x = \sqrt[3]{a + \sqrt{b}} + \sqrt[3]{a - \sqrt{b}}$$

Rather than using these more difficult formulas, find an estimate of a real root of the cubic equation

$$x^3 - 2a^2x - b^3 = 0$$

utilizing the properties of the Golden Ratio (*LM* p 54),

$$\varphi = \frac{1 + \sqrt{5}}{2} \approx 1.618$$

Note that powers of φ are equal to linear expressions of φ, that is $\varphi^2 = \varphi + 1$, $\varphi^3 = 2\varphi + 1$, etc.

4A Find an estimate for the equation: $x^3 + 2a^2x - b^3 = 0$

4B Explore the properties of Golden Ratio trapezoids.

Golden Ratio – Trapezoid Aesthetics

Trapezoid	Short Side	Long Side	Short Base	Long Base
Golden	s	φs	B	φb
Golden Right	s	φs	$\sqrt{\varphi^3}s$	$\sqrt{\varphi^5}s$
Short Base Trisosceles	s	s	s	φs
Long Base Trisosceles	s	s	s/φ	s
Half Golden Isosceles	s	s	$2s$	$2\varphi s$
Half Golden Right	s	$\sqrt{2}s$	φs	$\varphi^2 s$
Golden Isosceles	s	s	φs	$\varphi^2 s$

1. Make sketches of the Golden Ratio Trapezoids on the chart. Which is the most aesthetically appealing?
(Optional: Create a different golden ratio trapezoid of your own.)

2. For your selected trapezoid, calculate its angles, diagonals and area.

5 Maneuvering Around a Corner

A long rod, which must be kept level, needs to carried through a hallway, of width "*a*," which then makes a right-angle turn into a hallway of width "*b*."

Let α be the angle between the rod and the hallway wall, where the rod touches the hallway corner point (see diagram).

What is the length of the longest rod that can make it around that corner?

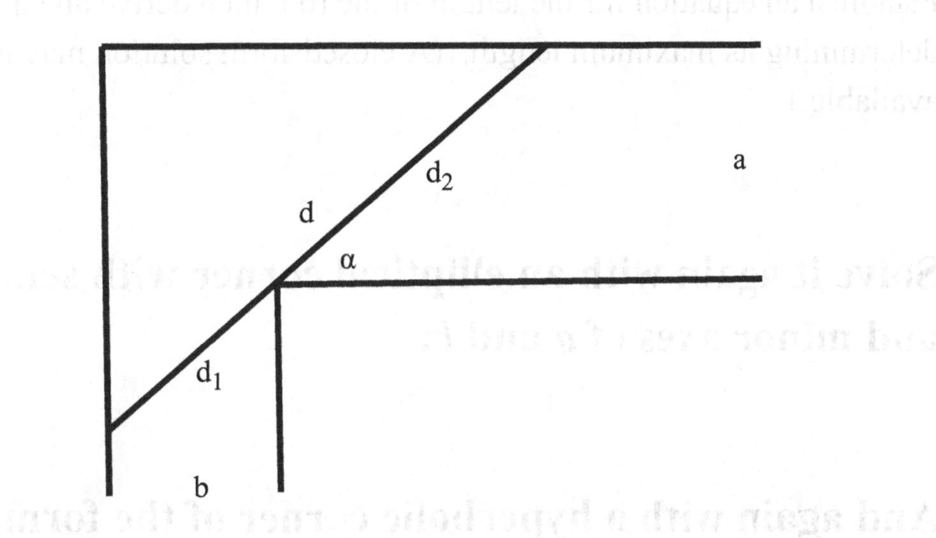

5A Solve the problem where the turning point is replaced with a circle quadrant of radius *c*.

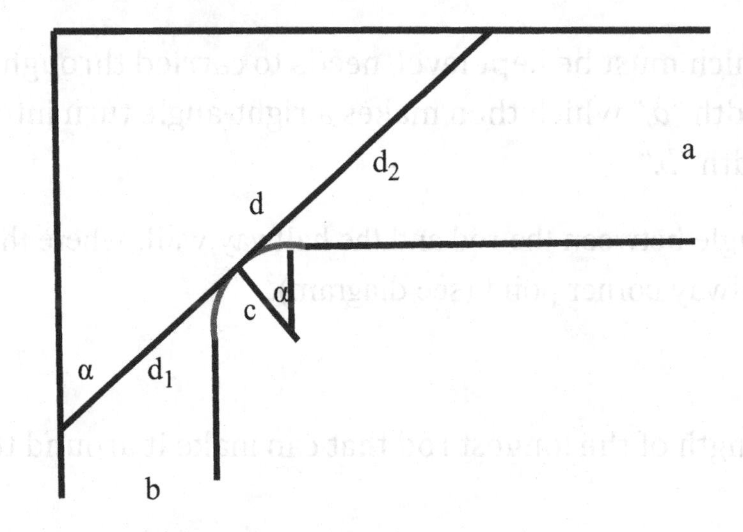

Establish an equation for the length of the rod; then derive an equation for determining its maximum length. (A closed-form solution may not be available.)

5B Solve it again with an elliptical corner with semi major and minor axes of *a* and *b*.

5C And again with a hyperbolic corner of the form $xy = c$.

5D And again with a parabolic corner of the form $4py = x^2$.

6 Calculus of Comparing Investment Performances

An investor makes two investments at the same time, $t_0 = 0$.

The first investment is for a_1 dollars and has a compound annual growth factor of b_1. That is, the value of the investment over time is: $a_1 b_1^t$.

The second investment is for a_2 dollars with a compound annual growth factor of b_2.

The investor knows that $a_1 > a_2$ and $b_1 < b_2$.

Let $y = a_1(b_1^{t_2}) - a_2(b_2^{t_2})$ be the difference in value of the two.

a) At what time t_2 will the two investments be worth the same?

b) What is a formula for t_1, the time when *y(t)* will be maximized?

c) Under what circumstances will the initial difference grow so that its maximum will occur at t_1, between t_0 and t_2?

d) What is the difference in values at t_1?

7 Calculating Important Parameters of an Investment (Retirement) Account

This problem is a real-life situation that senior citizens face in managing their financial resources during retirement. However, few retirees address these questions mathematically themselves; often their financial consultant will help with software tools.

The scene opens with a married couple (filing taxes jointly) who have no income other than their Social Security and their IRA accounts, which contain annuities and non-annuity investments.

The fundamental question they face is whether they will be able to lead the lifestyle they choose (at a given dollar cost per year) and still not run out of money before they die.

This problem does not take into account the effect of inflation or changes in the financial markets. These can be addressed as a further study later.

The parameters we will use are:

I initial IRA total amount (including annuity and non-annuity investments)

A value of annuities

$I - A$ value of non-annuity investments at the beginning of year one

k_I income rate from non-annuity investments

k_A income rate from annuities

S gross Social Security income per year

C Medicare (Part B) withholding from Social Security = \$2K/year

D total federal income tax deductions, plus exemptions

M the desired amount for spendable income every year, after taxes

We are narrowing the problem to the circumstances where the couples'

taxable income is \$75.3K to \$151.9K/year
(25% marginal tax bracket)

taxes are filed jointly with spouse, and there are no other sources of income

taxable Medicare income is the maximum, 0.85S

For the 25% tax bracket, the annual tax for the ith year is calculated as:

t_i = 10.4 + 0.25(non-annuity income + annuity income + additional withdrawals of principal + taxable Social Security – Part B premiums and Deductions/exemptions – 75.3)

$$t_i = 10.4 + 0.25(k_I x_i + k_A A + y_i + 0.85S - C - D - 75.3)$$

Note that principal can only be withdrawn from the non-annuity investments.
The desired spendable amount is

M = non-annuity income + annuity income + non-annuity principal withdrawals + take home Social Security – taxes

$M = k_I x_i + k_A A + y_i + S - 2.0 - t_i$ where

x_i the amount of money in the account at the beginning of year i, so that when

y_i is the amount of money withdrawn from the account principal, in addition to the annuity and non-annuity income, which will provide the exact amount necessary to pay for

t_i federal income tax and

M the desired spending money per year, assumed to be a constant

What we are seeking to find are three formulas, as functions of i for:

x_i the amount of money in the account at the beginning of year i, so that when

y_i is the amount of money withdrawn from the non-annuity principal, in addition to the annuity and non-annuity income, which will provide the exact amount necessary to pay for

t_i federal income tax (and, of course, M, a constant).

Once those formulas are obtained, then find

n the number of years until the account principal, and x_i, reach zero.

V the value of A above which investing in annuities decreases n

k_M the rate of return on annuities required to maximize n for a given M

8 Using Origami's Third Dimension to Double a Volume

Origami is the ancient Japanese art of folding paper into representative shapes. It is a marvelous confluence of art and science, and there are grand masters who can accomplish great feats with it.

Looking back into antiquity, there have been three problems that could not be solved using the classic geometric rules of construction with a straight edge and compass. These are the squaring of a circle, the trisection of an angle, and the doubling of a given volume. The squaring of a circle requires the construction of a ratio of lengths that is $\sqrt{\pi}$. Similarly, doubling a volume requires the construction of a ratio of lengths that is $\sqrt[3]{2}$.

Somehow, though, Origami, with its use of the third dimension that allows folding, enables the construction of a ratio of lengths of $\sqrt[3]{2}$ and makes doubling a volume possible.

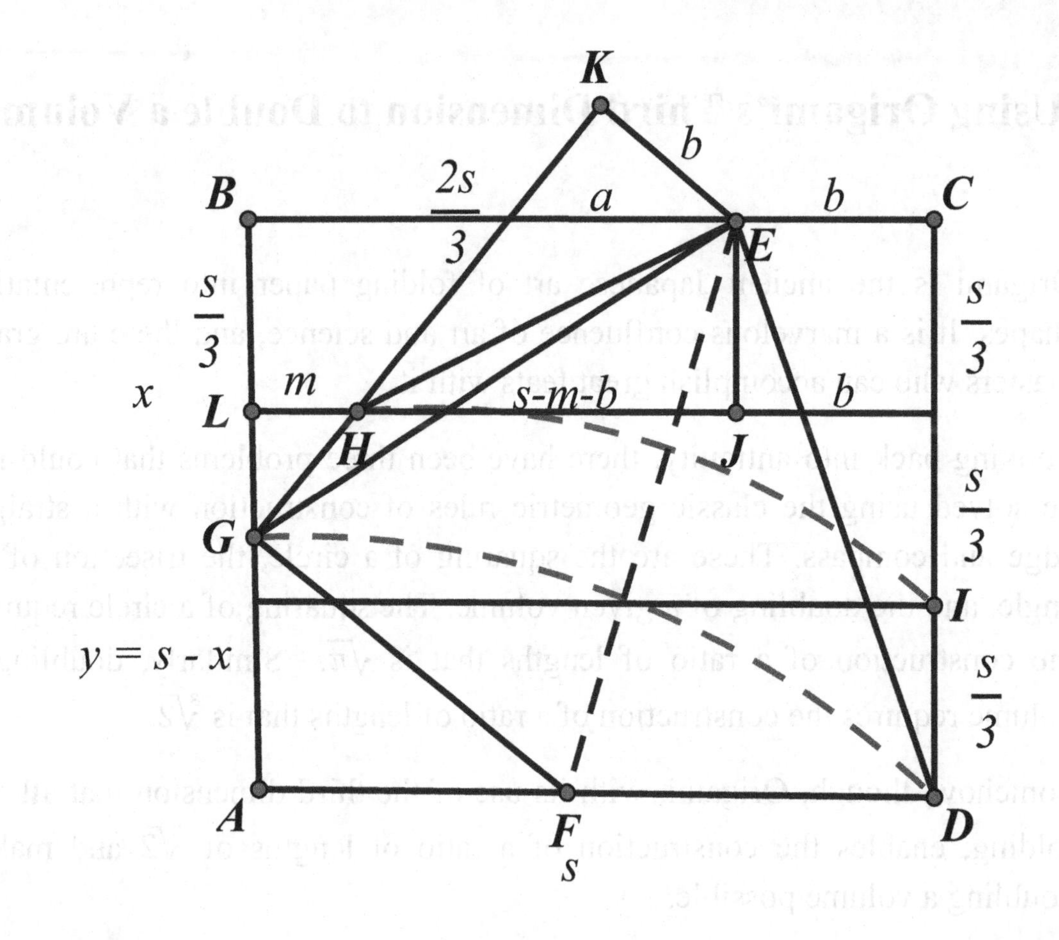

Given a square piece of paper *ABCD* of side *s*, locate points *E* and *F* such that point *D* folds onto *AB* at *G* and point *I* (which lies on one of the 1/3 folds) lies at point *H* (on the other 1/3 fold).

Given that *BG* has length *x* and *GA* has length $s - x$, show that

$$\frac{x}{s-x} = \sqrt[3]{2}$$

8A Explore whether Origami can be used to trisect an angle.

8B Explore whether Origami can be used to square a circle.

9 Solving for Equally Damped Sinusoids

A damped sinusoid is an oscillation that "dies out," such as the motion of a weight hanging on a spring. From the rest position, extend the string and watch the weight go up and down until it returns to the original position. Its equation of motion, that relates the height of the weight, *h(t),* and time, is:

$$h(t) = h_0 e^{-mt} \cos(nt) = h_0 Re[e^{-mt+nti}]$$

$$\text{or} \quad h(t) = h_0 e^{st} \text{ where } s = -m + ni$$

and h_0 is the weight's initial displacement. The graphs below give examples of the different ways such oscillations can die out based on damping.

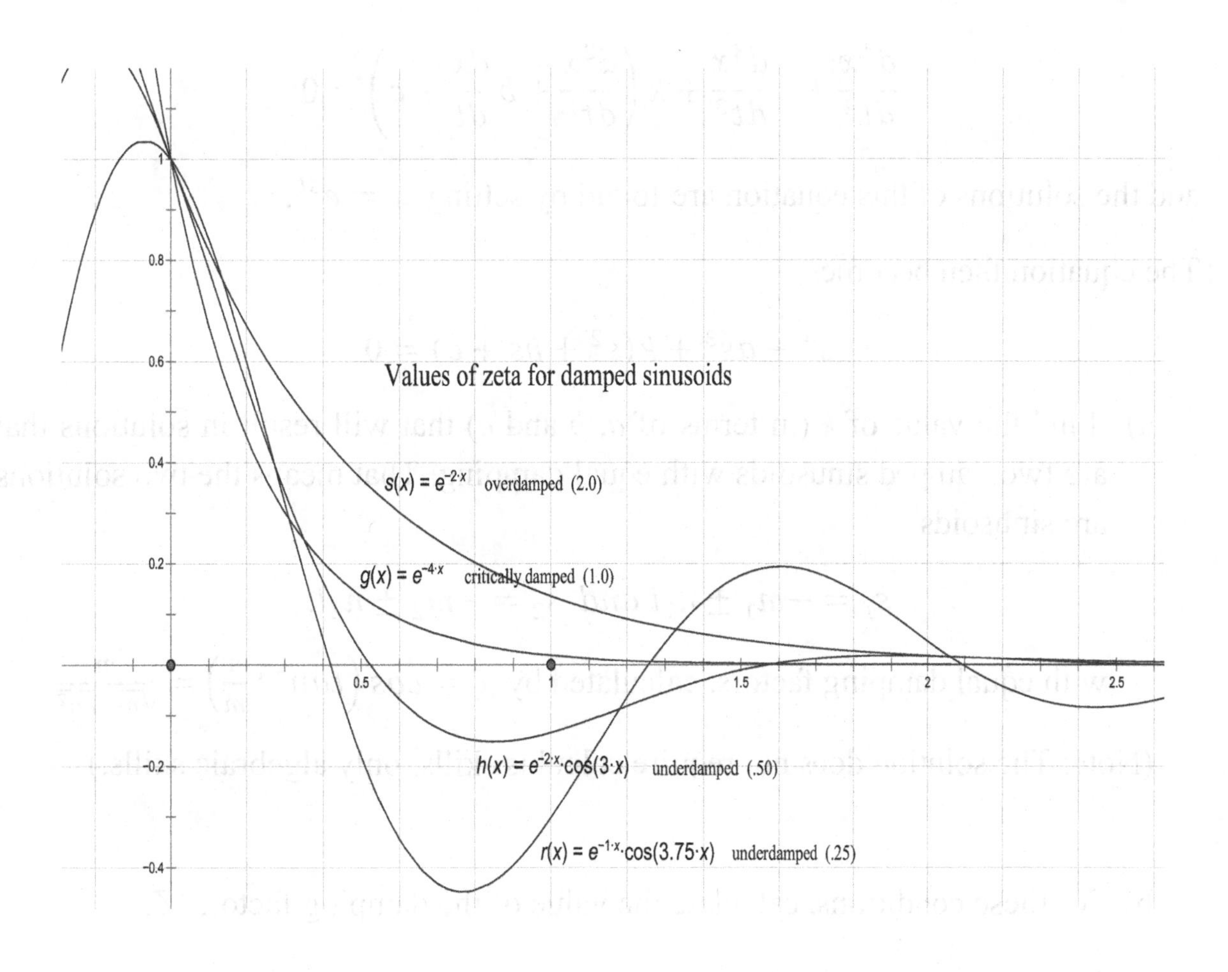

Informally, damping is a measure of how long it takes a sinusoid (sine wave or cosine wave) to reach and stay within a desired percentage of its initial or final value.

More formally, the damping factor (or damping ratio), symbolized by the Greek letter ζ, zeta, is an element of the solution

$$x(t) = e^{-\zeta\omega_0 t} cos\omega_0\sqrt{1-\zeta^2 t}$$

to a second-order differential equation

$$\frac{d^2x}{dt^2} + 2\zeta\omega_0\frac{dx}{dt} + \omega_0^2(1-\zeta^2)x=0.$$

ω_0 is the undamped frequency (or the frequency when $\zeta = 0$).

Applying this to the foundational differential equation of automatic control system theory:

$$\frac{d^4x}{dt^4} + a\frac{d^3x}{dt^3} + k\left(\frac{d^2x}{dt^2} + b\frac{dx}{dt} + c\right) = 0$$

and the solutions of this equation are found by setting $x = e^{st}$.

The equation then becomes

$$s^4 + as^3 + k(s^2 + bs + c) = 0$$

a) Find the value of *k* (in terms of *a*, *b* and *c*) that will result in solutions that are two damped sinusoids with equal damping. That means the two solutions are sinusoids

$$s_1 = -m_1 \pm n_1 i \ and \ s_2 = -m_2 \pm n_2 i,$$

with equal damping factors, calculated by $\zeta = cos\left(tan^{-1}\frac{n}{m}\right) = \frac{m}{\sqrt{m^2+n^2}}$

(Note: The solution does not require calculus skills, only algebraic skills.)

b) For these conditions, calculate the value of the damping factor, ζ.

11 Human Modeling Using Imaginary Numbers

We are accustomed to modeling tangible objects (like ourselves) in three physical and one time dimension. This enables us to locate ourselves in space, as well as determine our velocity and acceleration. But, there is more. How do we model the intangible, nonphysical parts of us? They are important, too.

Consider using imaginary numbers to form three mutually perpendicular axes. One axis could measure our logical thoughts, as in the left brain. One axis could measure our feelings, such as in our right brain. And one axis can be for our values, and this axis could be vertical (such as having higher aspirations or lower morals, etc.). Imaginary time is the last axis and has been used by mathematicians such as Minkowski, working with relativity, and physicists such as Hawking, working on the big bang theory.

What equations would result from rotating one of the three intangible/imaginary axes such as the x_i axis of values toward the t_i imaginary time axis?

And, what would it mean, on the human level?

Section II

Solutions, Discussions and Applications

Problems 1 - 10

1 Equal Angle Bisectors Theorem Revisited

Prove that if two angle bisectors of a triangle are equal,

then the triangle is isosceles.

See *LM* p 96 for the classical Steiner-Lehmus proof.

The objective now is to find another proof.

Hint: Use Stewart's Formula for angle bisectors. *LM* p 90.

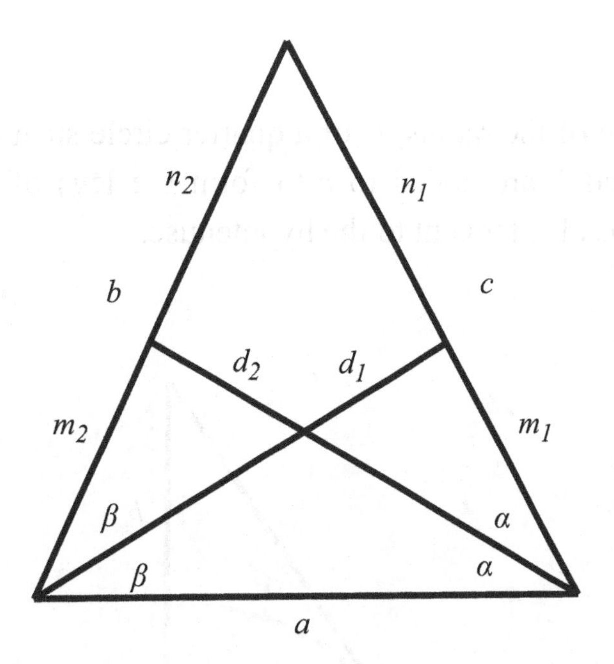

Using Stewart's Formula for angle bisectors

$$d_1^2 + m_1 n_1 = ab$$

$$d_2^2 + m_2 n_2 = ac$$

Since $d_1 = d_2$, given, $m_1 n_1 - m_2 n_2 = ab - ac$

Assume $\alpha > \beta$, then $m_2 > m_1$ and $n_2 > n_1$

So, $m_1 n_1 - m_2 n_2 < 0$, $ab - ac < 0$ and $b < c$.

This is a contradiction.

A similar contradiction occurs for $\alpha < \beta$.

Therefore $\alpha = \beta$ and $b = c$. QED

This proof is included because alternate solutions and approaches to a given problem are highly valued. (*LM* p 7)

2 Right Triangle with Inscribed Quarter Circle

Find the length of the radius, r, of a quarter circle such that when given line segments a and b are added to r to form the legs of a right triangle, the quarter circle is also tangent to the hypotenuse.

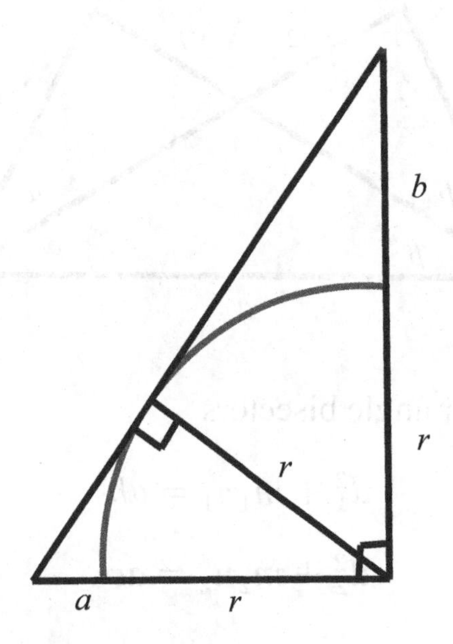

Using the area formulas,

$$r\sqrt{(a+r)^2+(b+r)^2}=(a+r)(b+r)$$

$$r^4=4abr^2+2ab(a+b)r+a^2\,b^2$$

Ferrari Method

$$m=4ab \qquad n=2ab(a+b) \qquad p=a^2\,b^2$$

$$r^4=mr^2+nr+p$$

$$m=m_1+m_2$$

$$r^4-m_1r^2=m_2r^2+nr+p$$

$$r^4 - m_1 r^2 + \left(\frac{m_1}{2}\right)^2 = m_2\left[r^2 + \left(\frac{n}{m_2}\right)r + \frac{p}{m_2} + \frac{1}{m_2}\left(\frac{m_1}{2}\right)^2\right]$$

$$\left(r^2 - \frac{m_1}{2}\right)^2 = m_2\left(r^2 - \frac{n}{2\ m_2}\right)^2 \text{ if } \frac{p}{m_2} + \frac{1}{m_2}\left(\frac{m_1}{2}\right)^2 = \left(\frac{n}{m_2}\right)^2$$

Solving for m_2 yields $m_2 = \frac{n^2}{4p+m_1^2} = m - m_1$

$$m_1^3 - mm_1^2 + 4pm_1 + (n^2 - 4mp) = 0$$

To suppress this cubic equation, set $m_1 = x + m/3$ then

$$x^3 + \alpha x + \beta = 0 \text{ where}$$

$$\alpha = \frac{1}{3}(12p - m^2) \qquad \beta = \frac{1}{27}(27n^2 - 2m^3 - 72mp)$$

Substituting for m_1, n and p

$$\alpha = -\frac{4}{3}a^2b^2 \qquad \beta = \frac{4a^2b^2}{27}(27a^2 + 27b^2 - 50ab)$$

Using the standard formula for the roots of a cubic equation

$$x = \frac{(ab)^{2/3}}{3}\left[\sqrt[3]{2(u+\sqrt{v})} + \sqrt[3]{2(u-\sqrt{v})}\right]$$

Where $u = 50ab - 27a^2 - 27b^2 \qquad v = u^2 - 16a^2b^2$

Returning to the equation for r

$$\left(r^2 - \frac{m_1}{2}\right)^2 = m_2\left(r + \frac{n}{2m_2}\right)^2$$

yields the quadratic $\qquad 2\sqrt{m_2}r^2 - 2m_2 r - (m_1\sqrt{m_2} + n) = 0)$

The closed-form solution to this problem is:

$$r = \frac{m_2 + \sqrt{m_2^2 + 2m_1m_2 + 4ab(a+b)\sqrt{m_2}}}{2\sqrt{m_2}}$$

Where

$$m_1 = \frac{4ab}{3} + \frac{(ab)^{2/3}}{2}\left[\sqrt[3]{2(u+\sqrt{v})} + \sqrt[3]{2(u-\sqrt{v})}\right]$$

$$m_2 = \frac{8ab}{3} - \frac{(ab)^{2/3}}{2}\left[\sqrt[3]{2(u+\sqrt{v})} + \sqrt[3]{2(u-\sqrt{v})}\right]$$

$$u = 50ab - 27a^2 - 27b^2 \qquad v = u^2 - 16a^2b^2$$

Corollary:

For the case where $a = b,$ we get

$$r\sqrt{(a+r)^2 + (a+r)^2} = (a+r)(a+r)$$

$$r^4 = 4a^2r^2 + 4a^3r + a^4$$

For this situation, we do not need to use the Ferrari method.

$$r^4 = a^2(2r+a)^2$$

$$r^2 = a(2r+a)$$

$$r = a(1+\sqrt{2})$$

Example: Let $a = 1 \quad b = 2$

$$u = 50ab - 27a^2 - 27b^2 = -35$$

$$v = u^2 - 16a^2b^2 = 1161$$

$$\sqrt{v} = 34.0735$$

$$u + \sqrt{v} = -.9265$$

$$u - \sqrt{v} = -69.0735$$

$$m_1 = \frac{4ab}{3} + \frac{(ab)^{2/3}}{2}\left[\sqrt[3]{2(u+\sqrt{v})} + \sqrt[3]{2(u-\sqrt{v})}\right] = -.7186$$

$$m_2 = \frac{8ab}{3} - \frac{(ab)^{2/3}}{2}\left[\sqrt[3]{2(u+\sqrt{v})} + \sqrt[3]{2(u-\sqrt{v})}\right] = 8.7186$$

$$r = \frac{m_2 + \sqrt{m_2^2 + 2m_1m_2 + 4ab(a+b)\sqrt{m_2}}}{2\sqrt{m_2}} = 3.4391$$

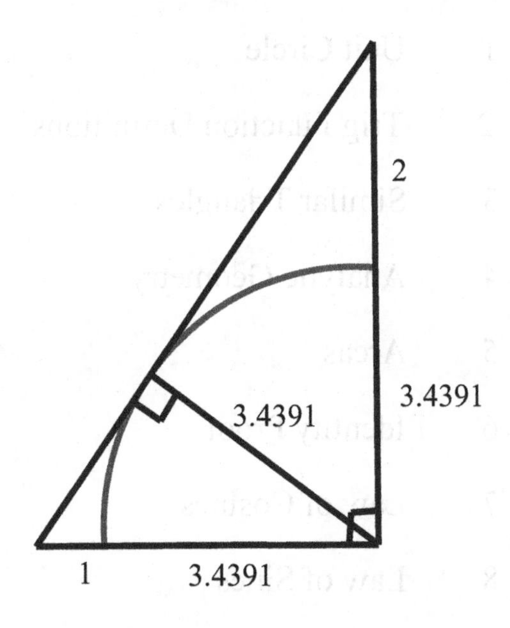

Check: $\frac{(3.4391+1)(3.4391+2)}{\sqrt{(3.4391+1)^2+(3.4391+2)^2}} = 3.4391$ QED

3 Ten Proofs of the Cosine of an Angle Sum

Present 10 proofs of the trigonometric identity:

$$cos(\alpha+\beta) = cos\,\alpha\,cos\,\beta - sin\,\alpha\,sin\,\beta$$

The pedagogy of finding more than one solution to a problem is highly valued. Here, 10 different approaches are used to prove one of the most useful trigonometric identities. Some of these solutions would normally flow from a trigonometry curriculum in their natural context. Other approaches may be studied later in the course or after trigonometry entirely. So some approaches are not available to the student when first studying the formula. Still, this is a good exercise for students to relate current and future concepts to previous ones.

There are 10 proofs shown here and, clearly, there are more. Students are encouraged to not only use these suggested approaches but also to develop others as well. The 10 presented here are:

1 Unit Circle

2 Trig Function Definitions

3 Similar Triangles

4 Analytic Geometry

5 Areas

6 Identity Proof

7 Law of Cosines

8 Law of Sines

9 Euler's Formula

10 Maclaurin Series

3 Ten Proofs of the Cosine of an Angle Sum

1 Unit Circle

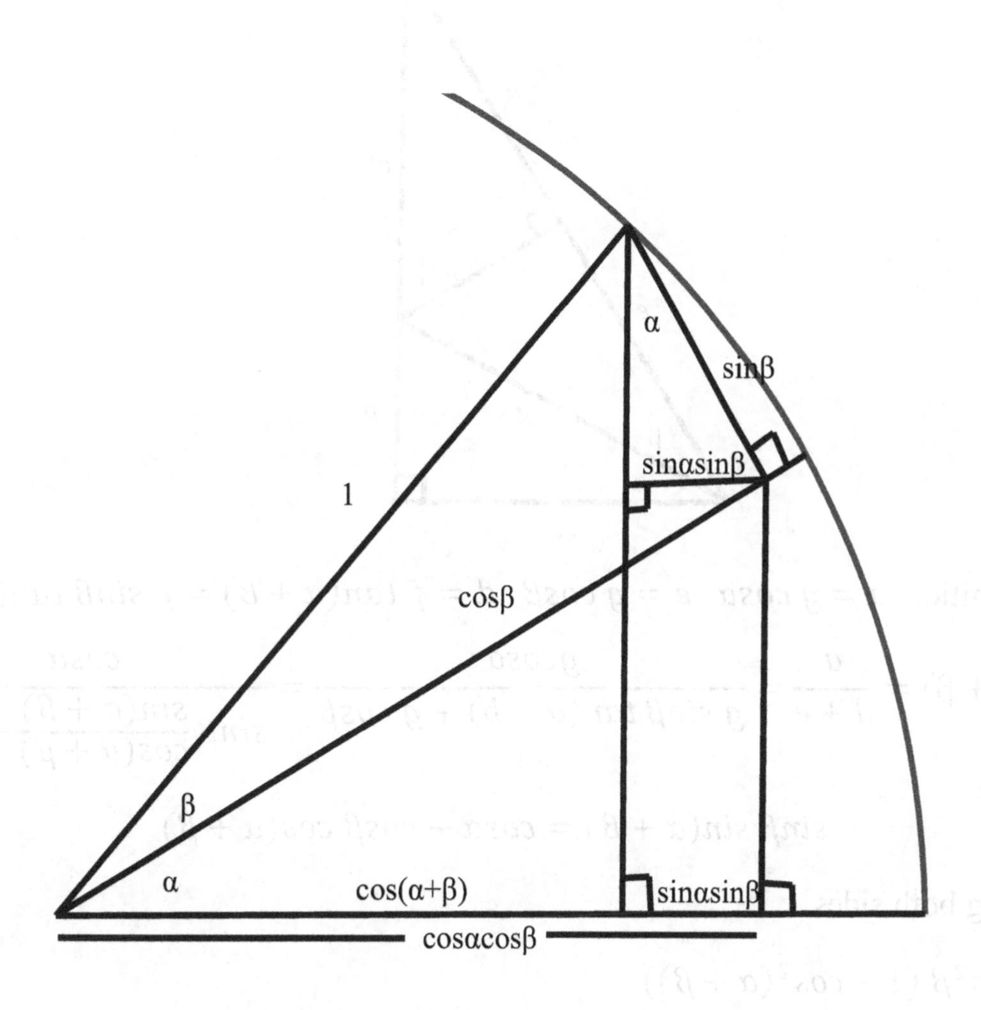

$$\cos(\alpha+\beta)=\cos\alpha\cos\beta-\sin\alpha\sin\beta$$

This diagram does all the talking — a picture is worth a thousand words.

3 Ten Proofs of the Cosine of an Angle Sum

2 Trigonometric Function Definitions

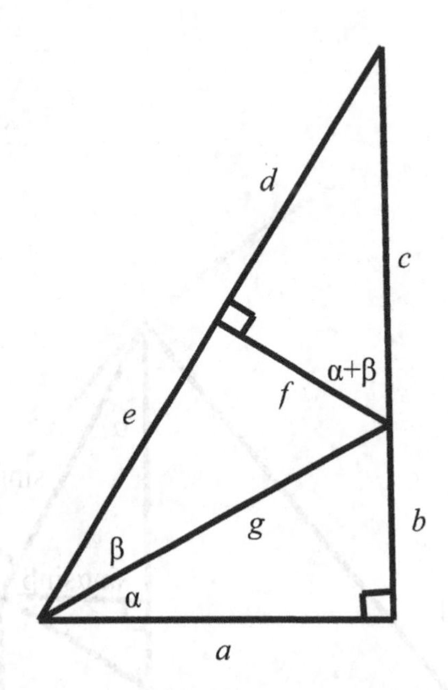

By definition, $a = g\ cos\alpha \quad e = g\ cos\beta \quad d = f\ tan(\alpha + \beta) = g\ sin\beta\ tan(\alpha + b)$

$$cos(\alpha + \beta) = \frac{a}{d + e} = \frac{gcos\alpha}{g\ sin\beta\ tan(\alpha + b) + g\ cos\beta} = \frac{cos\alpha}{sin\beta \dfrac{sin(\alpha + \beta)}{cos(\alpha + \beta)} + cos\beta}$$

$$sin\beta\ sin(\alpha + \beta) = cos\alpha - cos\beta\ cos(\alpha + \beta)$$

Squaring both sides,

$$sin^2\beta\ (1 - cos^2(\alpha + \beta)) \\ = cos^2\alpha - 2cos\alpha\ cos\beta\ cos(\alpha + \beta) + cos^2\beta\ cos^2(\alpha + \beta)$$

$$cos^2(\alpha + \beta) - 2cos\alpha\ cos\beta\ cos(\alpha + \beta) + (cos^2\alpha - sin^2\beta) = 0$$

$$cos(\alpha + \beta) = \frac{2cos\alpha\ cos\beta \pm \sqrt{4cos^2\alpha\ cos^2\beta - 4cos^2\alpha + 4sin^2\beta}}{2}$$

$$cos(\alpha + \beta) = cos\ \alpha\ cos\ \beta - sin\ \alpha\ sin\ \beta$$

3 Ten Proofs of the Cosine of an Angle Sum

3 Similar Triangles

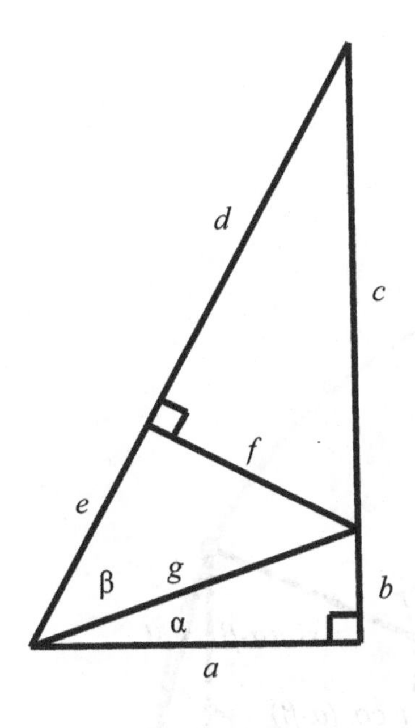

$$\frac{d}{f}=\frac{b+c}{a} \qquad d=\frac{f(b+c)}{a}$$

$$\frac{c}{f}=\frac{d+e}{a} \qquad c=\frac{f(d+e)}{a}$$

$$d=\frac{f\left(b+\frac{f(d+e)}{a}\right)}{a}=\frac{abf+ef^2}{a^2-f^2}$$

$$cos(\alpha+b)=\frac{a}{d+e}=\frac{a}{\frac{abf+ef^2}{a^2-f^2}+e}=\frac{a^2-f^2}{bf+ae}$$

$$a=gcos\alpha \quad b=gsin\alpha \quad e=gcos\beta \quad f=gsin\beta$$

$$cos(\alpha+\beta)=\frac{cos^2\alpha-sin^2\beta}{cos\alpha\,cos\beta+sin\alpha\,sin\beta}$$

$$=\frac{cos^2\alpha(cos^2\beta+sin^2\beta)-sin^2\beta(cos^2\alpha+sin^2\alpha)}{cos\alpha\,cos\beta+sin\alpha\,sin\beta}$$

$$=\frac{cos^2\alpha\,cos^2\beta-sin^2\alpha\,sin^2\beta}{cos\alpha\,cos\beta+sin\alpha\,sin\beta}$$

$$cos(\alpha+\beta)=cos\,\alpha\,cos\,\beta-sin\,\alpha\,sin\,\beta$$

3 Ten Proofs of the Cosine of an Angle Sum

4 Analytic Geometry

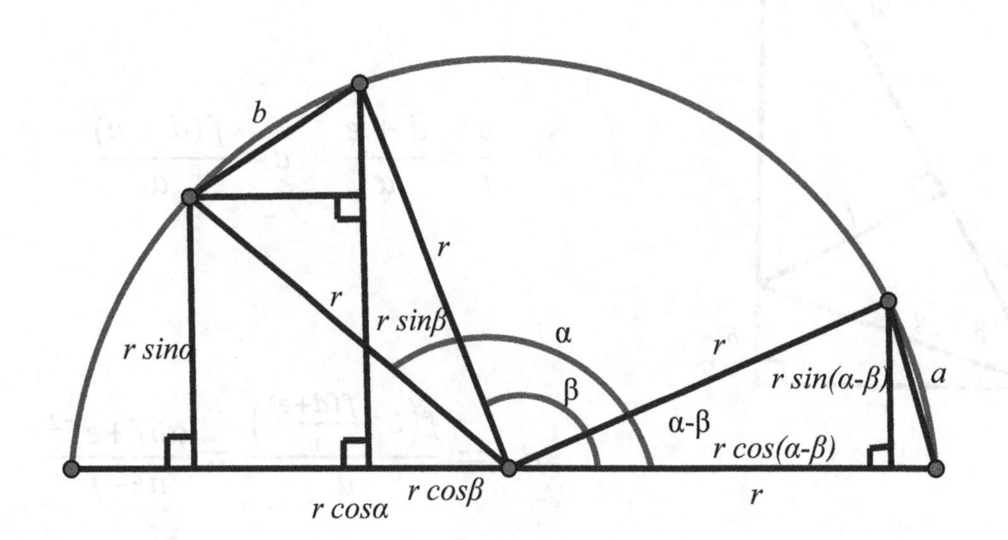

$$b = a$$

$$b^2 = a^2$$

$$(r\,cos\alpha - r\,cos\beta)^2 + (r\,sin\alpha - r\,sin\beta)^2$$
$$= [r\,cos(\alpha - \beta) - r]^2 + [r\,sin(\alpha - \beta)]^2$$

$$cos^2\alpha + sin^2\alpha + cos^2\beta + sin^2\beta - 2(cos\alpha\,cos\beta + sin\alpha\,sin\beta)$$
$$= cos^2(\alpha - \beta) + sin^2(\alpha - \beta) - 2\,cos(\alpha - \beta) + 1$$

$$cos(\alpha - \beta) = cos\,\alpha\,cos\,\beta + sin\,\alpha\,sin\,\beta$$

$$cos(\alpha + \beta) = cos\,\alpha\,cos\,\beta - sin\,\alpha\,sin\,\beta$$

3 Ten Proofs of the Cosine of an Angle Sum

5 Areas

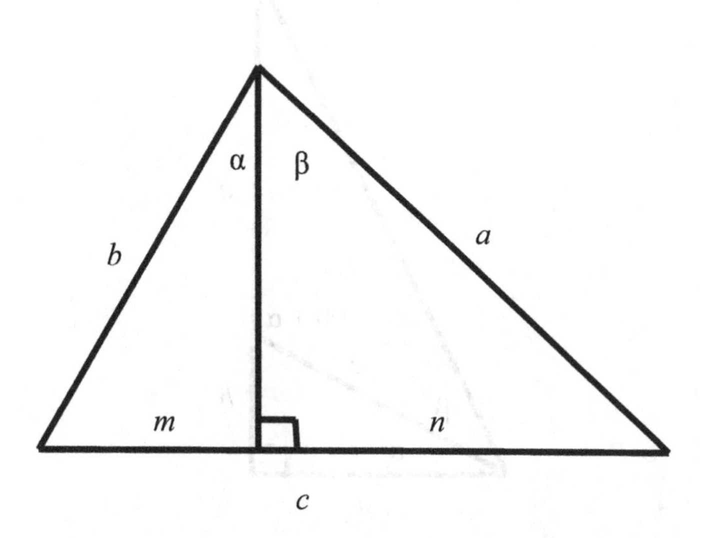

$$\frac{1}{2}ab\,sin(\alpha+\beta)=\frac{1}{2}ab\,sin\alpha cos\beta+\frac{1}{2}ab\,cos\alpha\,sin\beta$$

$$sin(\alpha+\beta)=sin\alpha\,cos\beta+cos\alpha\,sin\beta$$

$$cos^2(\alpha+\beta)=1-(sin\alpha\,cos\beta+cos\alpha\,sin\beta)^2=(cos\,\alpha\,cos\,\beta-sin\,\alpha\,sin\,\beta)^2$$

$$cos(\alpha+\beta)=cos\,\alpha\,cos\,\beta-sin\,\alpha\,sin\,\beta$$

6 Identity Proof

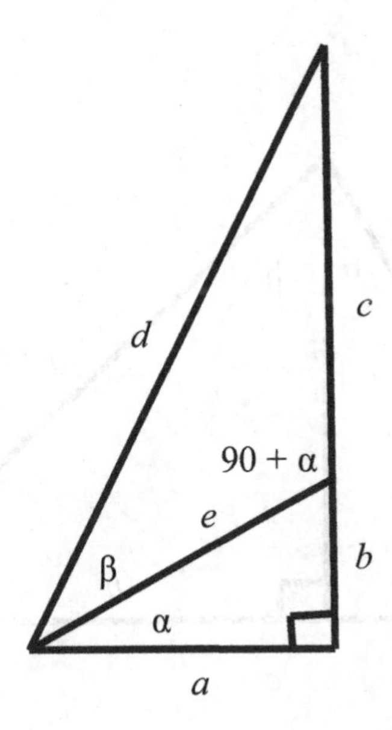

$\cos(\alpha+\beta) = \frac{a}{d}$ $\quad cos\alpha = \frac{a}{e}$ $\quad cos\beta = \frac{e^2+d^2-c^2}{2ed}$ Law of Cosines

$sin\alpha = \frac{b}{e}$ $\quad sin\beta = \frac{c}{d}\sin(90+\alpha) = \frac{c}{d}\cos\alpha = \frac{ca}{de}$ Law of Sines

$$cos(\alpha+\beta) = cos\,\alpha\, cos\,\beta - sin\,\alpha\, sin\,\beta$$

$$\frac{a}{d} = \frac{a}{e}\frac{e^2+d^2-c^2}{2ed} - \frac{b}{e}\frac{ca}{de}$$

$$\frac{a}{d} = \frac{a}{d}\frac{e^2+d^2-c^2-2bc}{2e^2}$$

$$e^2 = a^2+b^2 \qquad d^2 = a^2+(b+c)^2$$

$$\frac{a}{d} = \frac{a}{d}\frac{a^2+b^2+a^2+(b+c)^2-c^2-2bc}{2(a^2+b^2)}$$

$$\frac{a}{d} = \frac{a}{d}\frac{2(a^2+b^2)}{2(a^2+b^2)} = \frac{a}{d}$$ QED

3 Ten Proofs of the Cosine of an Angle Sum

7 Law of Cosines

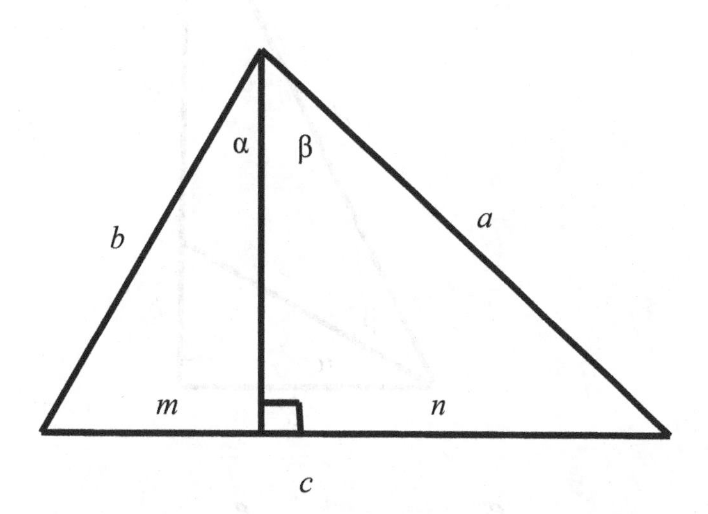

$$c^2 = (m+n)^2 = (m^2 + 2mn + n^2) = a^2 + b^2 - 2ab\,cos(\alpha + \beta)$$

$$m = b \sin \alpha \qquad n = a \sin \beta$$

$$b^2 sin^2\alpha + 2(b\,sin\alpha)(a\,sin\beta) + a^2 sin^2\beta = a^2 + b^2 - 2ab\,cos(\alpha + \beta)$$

$$2ab\,cos(\alpha + \beta) = a^2(1 - sin^2\beta) + b^2(1 - sin^2\alpha) - 2ab\,sin\alpha\,sin\beta$$

$$2ab\,cos(\alpha + \beta) = a^2\,(cos^2\beta) + b^2(cos^2\alpha) - 2ab\,sin\alpha\,sin\beta$$

$$a^2\,(cos^2\beta) = b^2(cos^2\alpha) = ab\,cos\alpha\,cos\beta$$

$$2ab\,cos(\alpha + \beta) = 2ab\,cos\alpha\,cos\beta - 2ab\,sin\alpha\,sin\beta$$

$$cos(\alpha + \beta) = cos\,\alpha\,cos\,\beta - sin\,\alpha\,sin\,\beta$$

3 Ten Proofs of the Cosine of an Angle Sum

8 Law of Sines

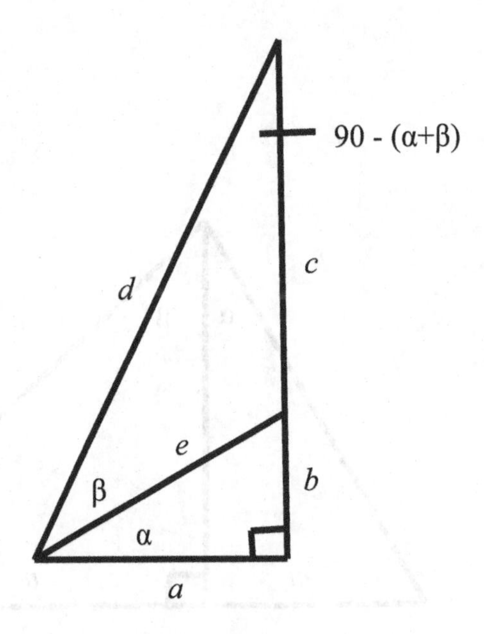

$$\frac{c}{sin\beta} = \frac{e}{sin[90-(\alpha+\beta)]} = \frac{e}{cos(\alpha+\beta)} \qquad c = \frac{e\ sin\beta}{cos(\alpha+\beta)}$$

$$cos(\alpha+\beta) = \frac{a}{\sqrt{a^2+(b+c)^2}} = \frac{e\ cos\alpha}{\sqrt{(e\ cos\alpha)^2 + (e\ sin\alpha + \frac{e\ sin\beta}{cos(\alpha+\beta)})^2}}$$

$$cos^2(\alpha+\beta) + 2\ sin\alpha\ sin\beta\ cos(\alpha+\beta) + (sin^2\beta - cos^2\alpha) = 0$$

$$cos(\alpha+\beta) = -sin\alpha\ sin\beta \pm \sqrt{sin^2\alpha\ sin^2\beta - sin^2\beta + cos^2\alpha}$$

$$sin^2\alpha\ sin^2\beta - sin^2\beta + cos^2\alpha = sin^2\beta\ (sin^2\alpha - 1) + cos^2\alpha =$$
$$sin^2\beta(-cos^2\alpha) + cos^2\alpha = cos^2\alpha\ (-sin^2\beta + 1) = cos^2\alpha\ cos^2\beta$$

$$cos(\alpha+\beta) = -sin\alpha\ sin\beta \pm \sqrt{cos^2\alpha\ cos^2\beta}$$

$$cos(\alpha+\beta) = cos\,\alpha\,cos\,\beta - sin\,\alpha\,sin\,\beta$$

3 Ten Proofs of the Cosine of an Angle Sum

9 Euler's Formula

$$cos(\alpha + \beta) = Re\left[e^{i(\alpha+\beta)}\right] = Re\left[e^{i\alpha}e^{i\beta}\right]$$

$$= Re[(cos\alpha + i\ sin\alpha)(cos\beta + i\ sin\beta)]$$

$$= Re[(cos\ \alpha\ cos\ \beta - sin\ \alpha\ sin\ \beta) + i(\ sin\ \alpha\ cos\ \beta + cos\ \alpha\ sin\ \beta)]$$

$$cos(\alpha + \beta) = cos\ \alpha\ cos\ \beta - sin\ \alpha\ sin\ \beta$$

3 Ten Proofs of the Cosine of an Angle Sum

10 Maclaurin Series

$$cos(\alpha+\beta) = 1 - \frac{(\alpha+\beta)^2}{2!} + \frac{(\alpha+\beta)^4}{4!} - \cdots$$

$$cos(\alpha) = 1 - \frac{\alpha^2}{2!} + \frac{\alpha^4}{4!} - \cdots$$

$$cos(\beta) = 1 - \frac{\beta^2}{2!} + \frac{\beta^4}{4!} - \cdots$$

$$sin(\alpha) = \alpha - \frac{\alpha^3}{3!} + \frac{\alpha^5}{5!} - \cdots$$

$$sin(\beta) = \beta - \frac{\beta^3}{3!} + \frac{\beta^5}{5!} - \cdots$$

$$cos\,a\ cos\beta = (1 - \frac{\alpha^2}{2!} + \frac{\alpha^4}{4!} - \cdots)(1 - \frac{\beta^2}{2!} + \frac{\beta^4}{4!} - \cdots)$$

$$= 1 - \frac{\alpha^2}{2!} + \frac{\alpha^4}{4!} - \frac{\beta^2}{2!} + \frac{\beta^4}{4!} + \frac{\alpha^2\beta^2}{2!2!} - \cdots$$

$$sin\alpha\ sin\beta = \left(\alpha - \frac{\alpha^3}{3!} + \frac{\alpha^5}{5!} - \cdots\right)\left(\beta - \frac{\beta^3}{3!} + \frac{\beta^5}{5!} - \cdots)\right)$$

$$= \alpha\beta - \frac{\alpha\beta^3}{3!} + \frac{\alpha\beta^5}{5!} - \frac{\alpha^3\beta}{3!} + \frac{\alpha^5\beta}{5!} - \cdots$$

$$cos(\alpha+\beta) = cos\,\alpha\,cos\,\beta - sin\,\alpha\,sin\,\beta$$

4 Simple Estimate for a Cubic Equation Root Using the Golden Ratio

Given the cubic equation

$$y^3 + py^2 + qy + r = 0$$

We know that if we let $y = x - \frac{p}{3}$

then the "depressed" or "reduced" equation is

$$x^3 - 3cx - 2a = 0$$

where $c = \frac{p^2}{q} - \frac{q}{3}$, $a = \frac{pq}{6} - \frac{p^3}{27} - \frac{r}{2}$ and $b = a^2 - c^3$.

A real root of this equation is

$$x = \sqrt[3]{a + \sqrt{b}} + \sqrt[3]{a - \sqrt{b}}$$

Rather than using these more difficult formulas, find an estimate of a real root of the cubic equation

$$x^3 - 2a^2x - b^3 = 0$$

utilizing the properties of the Golden Ratio (LM p 54),

$$\varphi = \frac{1 + \sqrt{5}}{2} \approx 1.618$$

Note that powers of φ are equal to linear expressions of φ, that is

$$\varphi^2 = \varphi + 1, \quad \varphi^3 = 2\varphi + 1, \text{ etc.}$$

For $$x^3 - 2a^2x - b^3 = 0$$

Let $$x = \varphi k$$

then $$\varphi^3 k^3 - 2a^2\varphi k - b^3 = 0$$

but $$\varphi^3 = 2\varphi + 1$$

so $$\varphi = -\frac{k^3 - b^3}{2k(k^2 - a^2)} = -\frac{(k-b)(k+b+\frac{b^2}{k})}{2(k-a)(k+a)}$$

Let $$m = \frac{k+b+\frac{b^2}{k}}{k+a}$$

then $$\varphi = -\frac{m(k-b)}{2(k-a)}$$

Solving for *k* yields $$k = \frac{2a\varphi + mb}{2\varphi + m}$$

Take a first estimate to be $m = 1$

then $$x = k\varphi = \frac{2a\varphi + b}{\varphi^2} \approx \frac{5}{4}a + \frac{2}{5}b$$

Substituting $a = rb$ and $k = \frac{2a\varphi + b}{\varphi^3}$ into $m = \frac{k+b+\frac{b^2}{k}}{k+a}$

yields a second estimate for *m*, i.e. $$m = \frac{.433r^2 + .835r + .959}{r^2 + .443r + .0414}$$

We can evaluate this approximation for different values of *r*, see below.

r	$m = \frac{.433r^2 + .835r + .959}{r^2 + .443r + .0414}$	$x = k\varphi = \frac{2a\varphi + mb}{2\varphi + m}\varphi$	Or $x =$	Actual Solution $x =$
.00	23.26	$.122a + .877b$	$.877b$	$1.00b = b$
.25	5.566	$.595a + 1.0233b$	$4.69a$ or $1.17b$	$4.79a$ or $1.20b$
.50	1.676	$1.066a + .552b$	$2.17a$	$2.33a$
.75	1.954	$1.009a + .609b$	$1.821a$	$1.818a$
1.0	1.500	$1.106a + .607b$	$1.713a$	$1.618a = \varphi a$
2.0	0.885	$1.270a + .3475b$	$1.4438a$	$1.4445a$
>>1	0.433	$1.427a + .191b$	$\approx 1.43a$	$1.414a = \sqrt{2}a$

The results are very good; the Golden Ratio estimated results are very close to the actual solutions with the greatest inaccuracy at $r = 0$.

There are many algorithms that can be developed for this data.

Based on the arc of the Golden Ratio solutions, one such algorithm, offered here, for solving equations of the form

$$x^3 - 2a^2x - b^3 = 0$$

with $r = \frac{a}{b}$ is

$$x = (1 + 0.1r)b = b + .1a \quad \text{for} \quad 0 \le r \le 0.5$$

$$x = \left[0.3(2 - r)^2 + \sqrt{2}\right]a \quad \text{for} \quad 0.5 \le r \le 2$$

$$x = \sqrt{2}a \quad \text{for} \quad 2 \le r$$

Notes:

This algorithm works the same whether a is positive or negative.

If b is negative, then the positive real root becomes negative;

that is, we can change the formulas to include the factor of $sgn(b)$.

Examples:

a) $$x^3 - 18x - 1000 = 0$$

$$a = 3 \quad b = 10 \quad r = 0.3$$

Estimated solution: $x = (1 + .1r)b = 1.03(10) = 10.3$

Actual solution: $x = 10.6$

b) $$x^3 - 1.445x - 1 = 0$$

$$a = .85 \quad b = 1 \quad r = .85$$

Estimated solution: $x = [0.3(2 - r)^2 + \sqrt{2}]a = 1.539$

Actual solution: $x = 1.459$

c) $$x^3 - 1800x - 1000 = 0$$

$$a = 30 \quad b = 10 \quad r = 3.0$$

Estimated Solution; $x = \sqrt{2}a = 42.426$

Actual solution: 42.701

d) $$x^3 - 65x + 70 = 0$$

$$2a^2 = 65 \text{ so } a = 5.701$$

$$b^3 = -70 \text{ so } b = -4.121$$

For estimate purposes, we use the absolute value of r,

so that $$r = |a/b| = 5.701/4.121 = 1.383.$$

Using our algorithm for an estimated solution:

$$x = sgn(b)[0.3(2 - r)^2 + \sqrt{2}]a$$

$$= -[.3(2 - 1.383)^2 + 1.414]5.701 = -8.712$$

The actual solution to the equation is: $x = -8.555$ (error is 1.8%)

This presentation demonstrates that Golden Ratio principles can be used to obtain very accurate cubic solutions in a short time.

4A Find an estimate for the equation: $x^3 + 2a^2x - b^3 = 0$

A variant study would be to consider the equation:

$$x^3 + 2a^2x - b^3 = 0$$

Can a similar analysis approach work with this form?

4B Explore the properties of Golden Ratio trapezoids.

Golden Ratio – Trapezoid Aesthetics

Trapezoid	Short Side	Long Side	Short Base	Long Base
Golden	s	φs	b	φb
Golden Right	s	φs	$\sqrt{\varphi^3}s$	$\sqrt{\varphi^5}s$
Short Base Trisosceles	s	s	s	φs
Long Base Trisosceles	s	s	s/φ	s
Half Golden Isosceles	s	s	$2s$	$2\varphi s$
Half Golden Right	s	$\sqrt{2}s$	φs	$\varphi^2 s$
Golden Isosceles	s	s	φs	$\varphi^2 s$

1. Make sketches of the Golden Ratio Trapezoids on the chart. Which is the most aesthetically appealing?
   (Optional: Create a different golden ratio trapezoid of your own.)

2. For your selected trapezoid, calculate its angles, diagonals and area.

The number of trapezoids that can utilize the Golden Ratio is many, and each has its own aesthetic. Students have responded well to this study, and the calculations of angles, diagonals and areas are worthy of a good academic effort.

5 Maneuvering Around a Corner

A long rod, which must be kept level, needs to carried through a hallway, of width "*a*," which then makes a right angle turn into a hallway of width "*b*."

Let α be the angle between the rod and the hallway wall where the rod touches the hallway corner point (see diagram).

What is the length of the longest rod that can make it around that corner?

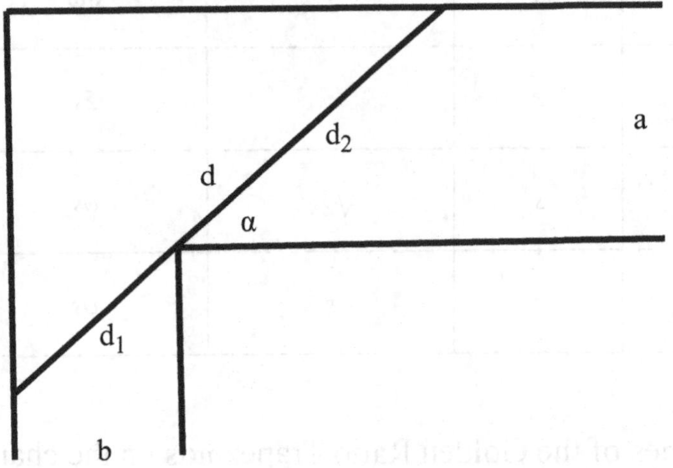

$$d_1 = b\ sec\alpha$$

$$d_2 = a\ csc\alpha$$

$$d = d_1 + d_2$$

$$d = a\ csc\alpha + b\ sec\alpha$$

To find the maximum length of the rod, we differentiate the rod length, *d*, with respect to α, and set it equal to zero.

$$d = a\, csc\alpha + b\, sec\alpha$$

$$\frac{dd}{d\alpha} = a(-cot\alpha\, csc\alpha) + b(tan\alpha\, sec\alpha) = 0$$

$$b\, sin^3\alpha = a\, cos^3\alpha \qquad tan^3\alpha = \frac{a}{b}$$

$$\alpha = tan^{-1}\sqrt[3]{\frac{a}{b}}$$

$$csc\alpha = \frac{(a^{2/3}+b^{2/3})^{1/2}}{a^{1/3}}$$

$$sec\alpha = \frac{(a^{2/3}+b^{2/3})^{1/2}}{b^{1/3}}$$

$$d = a\frac{(a^{2/3}+b^{2/3})^{1/2}}{a^{1/3}} + b\frac{(a^{2/3}+b^{2/3})^{1/2}}{b^{1/3}}$$

$$d = (a^{2/3}+b^{2/3})^{3/2}$$

Example: If $a = 10$ and $b = 7$, then

$$d = (10^{2/3}+7^{2/3})^{3/2} = 23.92$$

5A Solve the problem, where the turning point is replaced with a circle quadrant of radius *c*.

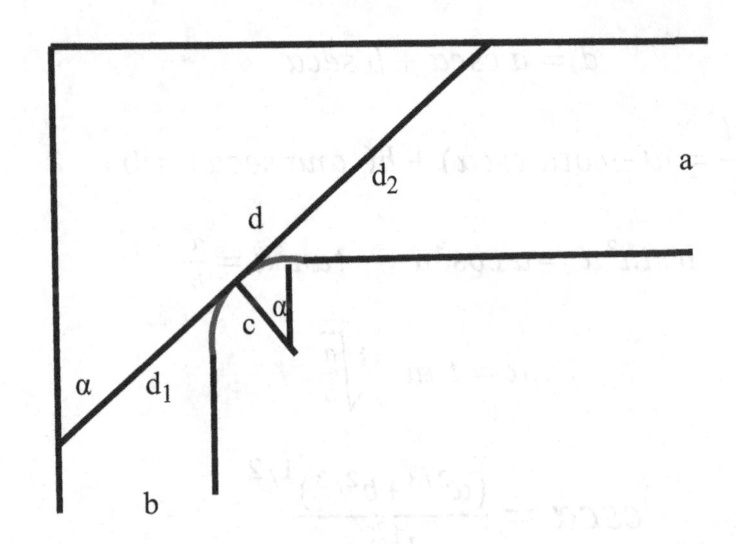

Establish an equation for the length of the rod; then derive an equation for determining its maximum length. (A closed-form solution may not be available.)

5B Solve it again with an elliptical corner with semi major and minor axes of *a* and *b*.

5C And again with a hyperbolic corner of the form $xy = c$.

5D And again with a parabolic corner of the form $4py = x^2$.

These problems become increasingly difficult and complicated. The solutions necessarily include the additional parameters of: a and b, c and p, respectively.

6 Calculus of Comparing Investment Performances

An investor makes two investments at the same time, $t_0 = 0$.

The first investment is for a_1 dollars, and it has a compound annual growth factor of b_1. That is, the value of the investment over time is: $a_1 b_1^t$.

The second investment is for a_2 dollars with a compound annual growth factor of b_2.

The investor knows that $a_1 > a_2$ and $b_1 < b_2$.

Let $y = a_1\left(b_1^{t_2}\right) - a_2\left(b_2^{t_2}\right)$ be the difference in value of the two.

a) At what time t_2 will the two investments be worth the same?

b) What is a formula for t_1, the time when *y(t)* will be maximized?

c) Under what circumstances will the initial difference grow so that its maximum will occur at t_1, between t_0 and t_2?

d) What is the difference in values at t_1?

a) The value of the investments will be the same when

$$y = a_1(b_1^{t_2}) - a_2(b_2^{t_2}) = 0$$

$$\left(\frac{b_1}{b_2}\right)^{t_2} = \frac{a_2}{a_1}$$

$$t_2 = \frac{ln\frac{a_2}{a_1}}{ln\frac{b_1}{b_2}}$$

b) To find the maximum value of y at t_1,

$$\frac{dy}{dt} = a_1(b_1^{t_1})(ln\, b_1) - a_2(b_2^{t_1})(ln\, b_2) = 0$$

$$\left(\frac{b_1}{b_2}\right)^{t_1} = \frac{a_2 ln\, b_2}{a_1 ln\, b_1}$$

$$t_1 = \frac{ln\left(\frac{a_2\, ln\, b_2}{a_1\, ln\, b_1}\right)}{ln\left(\frac{b_1}{b_2}\right)}$$

c) For a maximum at t_1 to exist,

$$ln\left(\frac{a_2\, ln\, b_2}{a_1\, ln\, b_1}\right) < 0 \text{ since } ln\left(\frac{b_1}{b_2}\right) < 0$$

$$\frac{a_2\, ln\, b_2}{a_1\, ln\, b_1} < 1$$

$$b_2 < b_1^{a_1/a_2}$$

d) The maximum difference in the values of the investments is

$$y = a_1 b_1 exp\left(\frac{ln\left(\frac{a_2\, ln\, b_2}{a_1\, ln\, b_1}\right)}{ln\left(\frac{b_1}{b_2}\right)}\right) - a_2 b_2 exp\left(\frac{ln\left(\frac{a_2\, ln\, b_2}{a_1\, ln\, b_1}\right)}{ln\left(\frac{b_1}{b_2}\right)}\right)$$

Example: $a_1 = 3.0 \quad a_2 = 1.5 \quad b_1 = 1.5 \quad b_2 = 2.0$

$$y = 3.0(1.5^t) - 1.5(2.0^t)$$

$$t_2 = \frac{ln\frac{a_2}{a_1}}{ln\frac{b_1}{b_2}} = \frac{ln(0.5)}{ln(.75)} = 2.409$$

A maximum exists because $2 < 1.5^{3/1.5} = 2.25$

and the value of *t* at the maximum is

$$t_1 = \frac{ln\left(\frac{a_2\, ln\, b_2}{a_1\, ln\, b_1}\right)}{ln\left(\frac{b_1}{b_2}\right)} = \frac{ln\left(\frac{1.5\, ln\, 2}{3\, ln\, 1.5}\right)}{ln\left(\frac{1.5}{2}\right)} = .546$$

The difference in the investment values is

$$y = 3(1.5)^{.546} - 1.5(2)^{.546} = 1.553 > 1.5$$

See graph below.

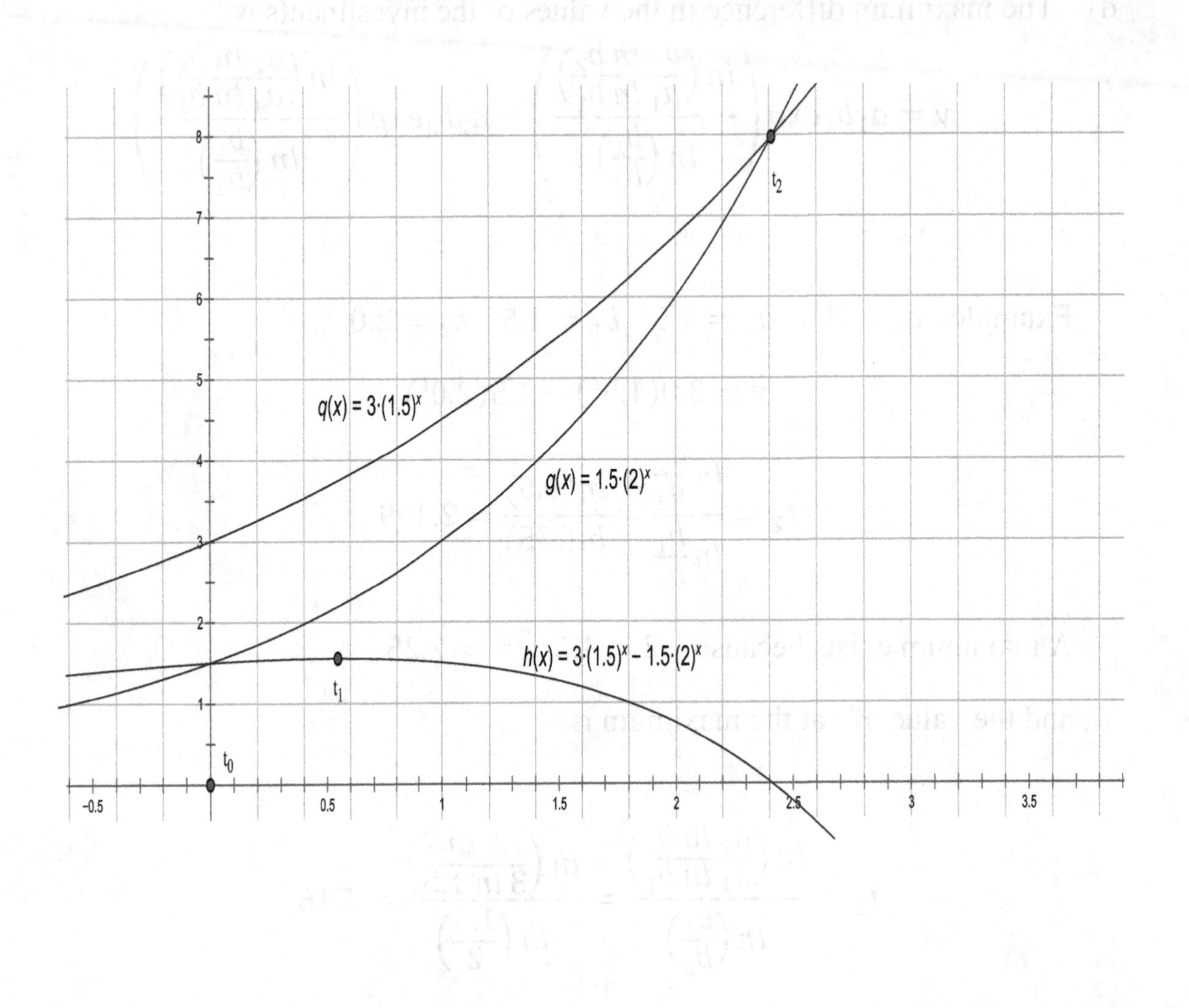

7 Calculating Important Parameters of an Investment (Retirement) Account

This problem is a real-life situation that senior citizens face in managing their financial resources during retirement However, few retirees address these questions mathematically themselves; often their financial consultant will help with software tools.

The scene opens with a married couple (filing taxes jointly) who have no income other than their Social Security and their IRA accounts, which contain annuities and non-annuity investments.

The fundamental question they face is whether they will be able to lead the lifestyle they choose (at a given dollar cost per year) and still not run out of money before they die.

This problem does not take into account the effect of inflation or changes in the financial markets. These can be addressed as a further study later.

The parameters we will use are:

I initial IRA total amount (including annuity and non-annuity investments.)

A value of annuities

$I - A$ value of non-annuity investments at the beginning of year one

k_I income rate from non-annuity investments

k_A income rate from annuities

S gross Social Security income per year

C Medicare (Part B) withholding from Social Security = $2K/year

D total federal income tax deductions, plus exemptions

M the desired amount for spendable income every year, after taxes

We are narrowing the problem to the circumstances where the couples'

taxable income is \$75.3K to \$151.9K/year
(25% marginal tax bracket)

taxes are filed jointly with spouse, and there are no other sources of income

taxable Medicare income is the maximum, 0.85S

For the 25% tax bracket, the annual tax for the ith year is calculated as:

t_i = 10.4 + 0.25(non-annuity income + annuity income + additional withdrawals of principal + taxable Social Security – Part B premiums and Deductions/exemptions – 75.3)

$$t_i = 10.4 + 0.25(k_I x_i + k_A A + y_i + 0.85S - C - D - 75.3)$$

Note that principal can only be withdrawn from the non-annuity investments. The desired spendable amount is

M = non-annuity income + annuity income + non-annuity principal withdrawals + take home Social Security - taxes

$M = k_I x_i + k_A A + y_i + S - 2.0 - t_i$ where

x_i the amount of money in the account at the beginning of year i, so that when

y_i is the amount of money withdrawn from the account principal, in addition to the annuity and non-annuity income, that will provide the exact amount necessary to pay for

t_i federal income tax and

M the desired spending money per year, assumed to be a constant

What we are seeking to find are three formulas, as functions of i for:

x_i the amount of money in the account at the beginning of year i, so that when

y_i is the amount of money withdrawn from the non-annuity principal, in addition to the annuity and non-annuity income, that will provide the exact amount necessary to pay for

t_i federal income tax (and, of course, M, a constant).

Once those formulas are obtained, then find

n the number of years until the account principal, and x_i, reach zero.

V the value of M which maximizes n

k_M the rate of return on annuities required to maximize n for a given M

Substituting t_i into *M*, combining like terms, and solving for y_i yields

$$y_i = 1.333M - k_I x_I - k_A A - 1.050S - 0.333D - 8.567 .$$

We can simplify the algebra by representing $y_i = \alpha x_i + \beta$

where $\alpha = -k_I$ and

$$\beta = 1.333M - k_A A - 1.050S - 0.333D - 8.567$$

At the beginning of the first year,

$$x_i = I - A - y_i$$

At the beginning of the second year,

$$x_2 = x_1 - y_2 = I - A - y_1 - y_2$$

And, at the beginning of the ith year,

$$x_i = I - A - \sum_{m=1}^{i} y_m$$

Substituting for the summation of y_m we get

$$x_i = I - A - \alpha \sum_{m=1}^{i} x_m - \beta i$$

and

$$x_{i-1} = I - A - \alpha \sum_{m=1}^{i-1} x_m - \beta(i-1)$$

Subtracting

$$x_i - x_{i-1} = -\alpha x_i - \beta$$

And

$$(1+\alpha)x_i = x_{i-1} - \beta$$

Solving for x_i ,

$$x_i = \frac{1}{1+\alpha}x_{i-1} - \frac{\beta}{1+\alpha}$$

Again, to simplify the algebra, let

$$x_i = ax_{i-1} + b$$

where $a = \frac{1}{1+\alpha}$ and $b = -\frac{\beta}{1+\alpha}$

We can find a recursive formula for these equations.

For $i = 1$, call $x_1 = c = \frac{I-A-\beta}{1+\alpha}$

Then

$$x_1 = c$$

$$x_2 = ac + b$$

$$x_3 = a(ac+b) + b = a^2c + b(1+a)$$

$$x_4 = a[a^2c + b(1+a)] + b = a^3c + b(1+a+a^2)$$

And, in general $x_i = a^{i-1}c + b(1 + a + a^2 + \cdots + a^{i-1})$

The terms $1 + a + a^2 + \cdots + a^{i-1}$

form a geometric series whose sum is $\frac{1-a^{i-1}}{1-a}$.

We have now found the solution for the amount of money in the account at the beginning of the ith year:

$$x_i = a^{i-1}c + b\left(\frac{1-a^{i-1}}{1-a}\right)$$

And we can calculate the amount of money to withdraw

$$y_i = \alpha x_i + \beta$$

and federal taxes to be paid

$$t_i = 10.4 + 0.25(k_I x_i + k_A A + y_i + 0.85S - D - 75.3) \ .$$

These calculations are made using our previous findings

$$\alpha = -k_I$$

$$\beta = 1.333M - k_A A - 1.050S - 0.333D - 8.567$$

$$a = \frac{1}{1+\alpha}$$

$$b = -\frac{\beta}{1+\alpha}$$

$$c = \frac{I - A - \beta}{1+\alpha}$$

Now, the most important parameter to calculate is

n the number of years until the account principal, x_i , reaches zero.

For this, we set $x_n = 0$.

$$x_n = a^{n-1}c + b\left(\frac{1-a^{n-1}}{1-a}\right) = 0$$

Multiplying by $(1 - a)$, and rearranging, yields

$$a^{n-1} = \frac{b}{b + c(a-1)}$$

And the solution is

$$n = 1 + \frac{\log\left[\frac{b}{b+c(a-1)}\right]}{\log a}$$

If we substitute for *a, b* and *c* and do algebraic manipulations, the equation

becomes: $$n = \frac{\log\left(1-k_I\frac{I-A}{\beta}\right)}{\log(1-k_I)}$$

The equation for n in this form allows us to understand additional properties of the IRA account. If

$$0 < \frac{I-A}{\beta} \leq 1 \text{ then } n < 0.$$

This is not a good investment design. If we wish to maximize n, then we need to maximize

$$\frac{I-A}{\beta} = \frac{I-A}{1.333M-k_A A-1.0505S-.333D-8.567}.$$

We can differentiate with respect to A and set equal to zero. The result is

$$(1.333M - k_A A - 1.0505S - .333D - 8.567)(-1) - (I - A)(-k_A) = 0$$

We find that

$$M = .7880S + 0.25D + 6.427 + 0.750k_A.$$

Another method of obtaining this result is to take two values of $\frac{I-A}{\beta}$ and set them equal to one another. That is:

$$\frac{I-A_1}{\beta_1} = \frac{I-A_2}{\beta_2} \text{ with}$$

$$\beta_1 = 1.333M - k_A A_1 - 1.050S - 0.333D - 8.567$$

$$\beta_2 = 1.333M - k_A A_2 - 1.050S - 0.333D - 8.567$$

Cross multiplying and solving for M yields

$$M = 0.7877S + 0.25D + 6.4269 + 0.75Ik_A$$

Let V be the value of M that maximizes n, then

$$V = 0.7877S + 0.25D + 6.4269 + 0.75Ik_A$$

This result is significant because the maximum value of n is independent of the amount invested in annuities, A.

If the amount invested in annuities is lower, the amount of money in nonannuity investments is higher but at a lower rate. To reach the same level of annual withdrawal, M, additional equity needs to be drawn. But, since the nonannuity equity is higher, the money will last longer. Similarly, if annuities are higher, less equity needs to be withdrawn, but there is less equity available. So the two cases come out even.

The equation indicates that once we have established values for S, D and I, which in real life are generally set in advance, the only true variable we can select in our investing is k_A , which is the rate of return our annuity provides. It is intuitively obvious that the greater rate we can get from our annuities, the greater amount we can take from the account annually while maximizing n, the length of time the account will last.

Lastly, for a given value of M, what is the required k_M to maximize n ?

We have:

$$M = 0.7877S + 0.25D + 6.4269 + 0.75Ik_M$$

so

$$k_M = \frac{0.7877S + 0.25D + 6.4269 - M}{.75I} = \frac{1.050S + .333D + 8.5692 - 1.333M}{I}$$

Example: (Monetary values given are in \$K.)

$$I = 900 \qquad A = 100 \qquad k_I = .06 \quad k_A = .08$$

$$S = 50 \qquad C = 2 \qquad D = 30 \quad M = 120$$

Find the value of the account at the beginning of the 8th year, the amount to be withdrawn that year, the amount in taxes for that year, and the number of years until the non-annuity funds have been depleted.

Solution:

$i = 8 \qquad\qquad \alpha = -k_I = -0.06$

$\beta = 1.333M - k_A A - 1.050S - 0.333D - 8.567 = 80.894$

$a = \frac{1}{1+\alpha} = 1.0638 \qquad b = -\frac{\beta}{1+\alpha} = -86.055$

$c = \frac{I - A - \beta}{1+\alpha} = 765.006 \qquad x_8 = a^{8-1}c + b\left(\frac{1-a^{8-1}}{1-a}\right) = 448.711$

$y_8 = \alpha x_8 + \beta = 53.971$

$t_8 = 10.4 + 0.25(k_I x_8 + k_A A + y_8 + 0.85S - D - 75.3) = 16.923$

$$n = 1 + \frac{\log\left[\frac{b}{b + c(a-1)}\right]}{\log a} = 14.54$$

Check:

$M = k_I x_8 + k_A A + y_8 + S - 2.0 - t_8 =$

$(.06)(448.711)+(.08)(100)+53.971+50-2-16.923=119.97 = M$ QED

$V = 0.7877S + 0.25D + 6.4269 + 0.75Ik_A = 107.31.$

Therefore, if M is reduced to 107.3, n would reach a maximum of 33.5 years

If $M = 110$, to maximize n,

$$k_M = \frac{1.050S + .333D + 8.5692 - 1.333M}{I} = \frac{1.050(50) + .333(30) + 8.5692 - 1.333(110)}{900} = .242$$

8 Using Origami's Third Dimension to Double a Volume

Origami is the ancient Japanese art of folding paper into representative shapes. It is a marvelous confluence of art and science, and there are grand masters who can accomplish great feats with it.

Looking back into antiquity, there have been three problems that could not be solved using the classic geometric rules of construction with a straight edge and compass. These are the squaring of a circle, the trisection of an angle, and the doubling of a given volume. The squaring of a circle requires the construction of a ratio of lengths that is $\sqrt{\pi}$. Similarly, doubling a volume requires the construction of a ratio of lengths that is $\sqrt[3]{2}$.

Somehow, though, Origami, with its use of the third dimension — which allows folding — enables the construction of a ratio of lengths of $\sqrt[3]{2}$ and makes doubling a volume possible.

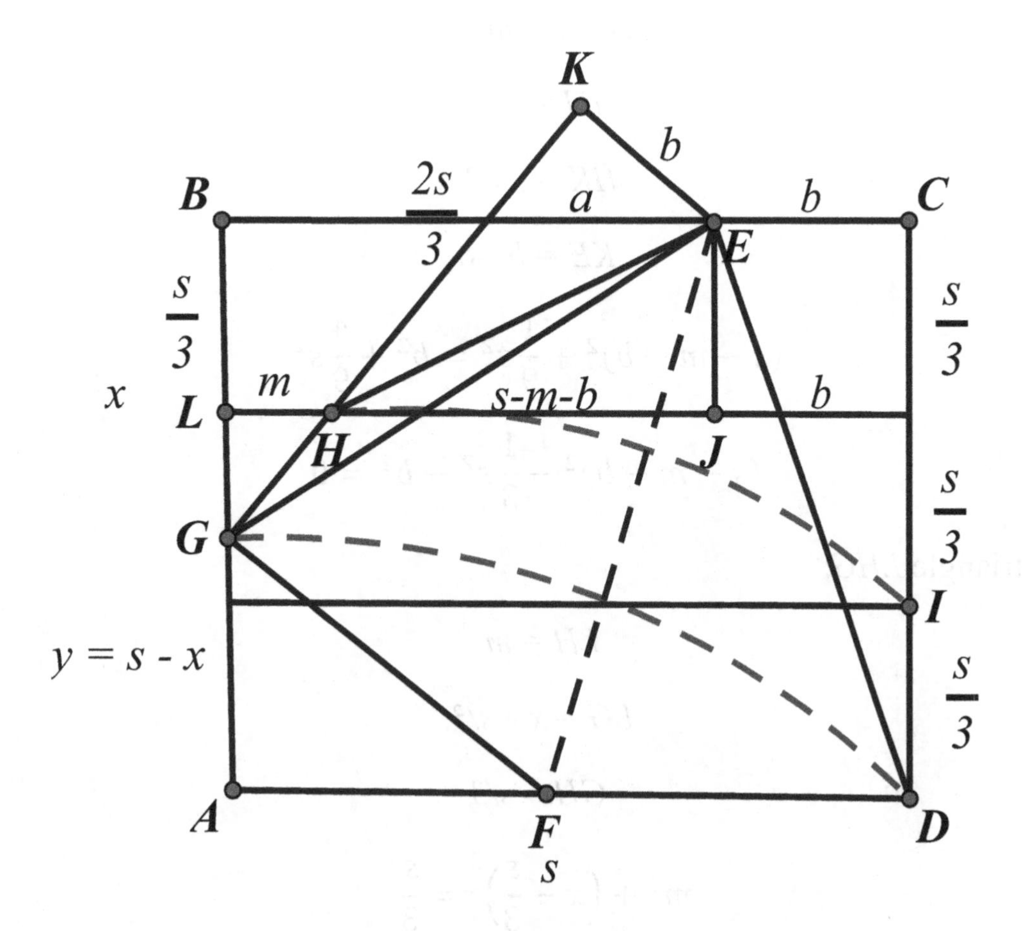

Given a square piece of paper *ABCD* of side *s*, locate points *E* and *F* such that point *D* folds onto *AB* at *G* and point *I* (which lies on one of the 1/3 folds) lies at point *H* (on the other 1/3 fold).

Given that *BG* has length *x* and *GA* has length *s* – *x*, show that

$$\frac{x}{s-x} = \sqrt[3]{2}$$

In right triangles *HJE* and *HKE*:

$$HJ = s - m - b,$$

$$EJ = s/3,$$

$$HK = 2s/3,$$

$$KE = b \text{ so}$$

$$(s-m-b)^2 + \frac{1}{9}s^2 = b^2 + \frac{4}{9}s^2$$

$$(s-m-b)^2 - \frac{1}{3}s^2 - b^2 = 0$$

In right triangle *LHG*:

$$LH = m,$$

$$LG = x - s/3,$$

$$GH = s/3$$

$$m^2 + \left(x - \frac{s}{3}\right)^2 = \frac{s^2}{3}$$

$$m = \sqrt{x\left(\frac{2s}{3} - x\right)}$$

In right triangles *BGE* and *CDE*:

$$BG = x, \quad BE = a, \quad CD = s,$$

$$CE = b \text{ and } \quad EG = ED, \text{ then}$$

$$x^2 + a^2 = s^2 + b^2 \text{ and } a = s - b \text{ so}$$

$$x^2 + (s-b)^2 = s^2 + b^2 \quad \text{and} \quad b = \frac{x^2}{2s}$$

Substituting gives:

$$\left(s - \sqrt{x\left(\frac{2s}{3} - x\right)} - \frac{x^2}{2s}\right)^2 - \frac{1}{3}s^2 - \left(\frac{x^2}{2s}\right)^2 = 0$$

After much expansion, the equation becomes

$$9x^6 - 6sx^5 + 16s^4x^4 - 16s^5x + 4s^6 = 0$$

This can be factored as

$$(3x^3 - 6sx^2 + 6s^2x - 2s^3)(3x^3 + 4sx^2 + 2s^2x - 2s^3) = 0$$

We can set the first factor equal to 0, and using the standard formula for the solution of a cubic equation,

$$3x^3 - 6sx^2 + 6s^2x - 2s^3 = 0$$

$$x = \frac{s}{3}(2 + \sqrt[3]{2} - \sqrt[3]{4})$$

Then $$s - x = \frac{s}{3}(1 - \sqrt[3]{2} + \sqrt[3]{4})$$

And the ratio is $$\frac{x}{s-x} = \frac{\frac{s}{3}(2+\sqrt[3]{2}-\sqrt[3]{4})}{\frac{s}{3}(1-\sqrt[3]{2}+\sqrt[3]{4})}$$

$$\frac{x}{s-x} = \sqrt[3]{2}\frac{(\sqrt[3]{4}+1-\sqrt[3]{2})}{(1-\sqrt[3]{2}+\sqrt[3]{4})} = \sqrt[3]{2} \quad \text{QED}$$

There is a key point here and that is that we can find a whole new approach to an old problem if we can gain the perspective of an extra dimension. What is difficult to see in a limited frame of reference becomes almost obvious when viewed from a new vantage point. This idea can also be a metaphor for addressing the challenges of human interaction and will be addressed further in Section III.

8A Explore whether Origami can be used to trisect an angle.

8B Explore whether Origami can be used to square a circle.

The student is encouraged to investigate the possibilities of the use of Origami in solving these ancient problems. With additional study, Origami becomes an extraordinary confluence of art, science and mathematics.

9 Solving for Equally Damped Sinusoids

A damped sinusoid is an oscillation that "dies out," such as the motion of a weight hanging on a spring. From the rest position, extend the string and watch the weight go up and down, until it returns to the original position. Its equation of motion, that relates the height of the weight, *h(t),* and time, is:

$$h(t) = h_0 e^{-mt} \cos(nt) = h_0 Re[e^{-mt+nti}]$$

$$\text{or} \quad h(t) = h_0 e^{st} \text{ where } s = -m + ni$$

and h_0 is the weight's initial displacement. The graphs below give examples of the different ways such oscillations can die out based on damping.

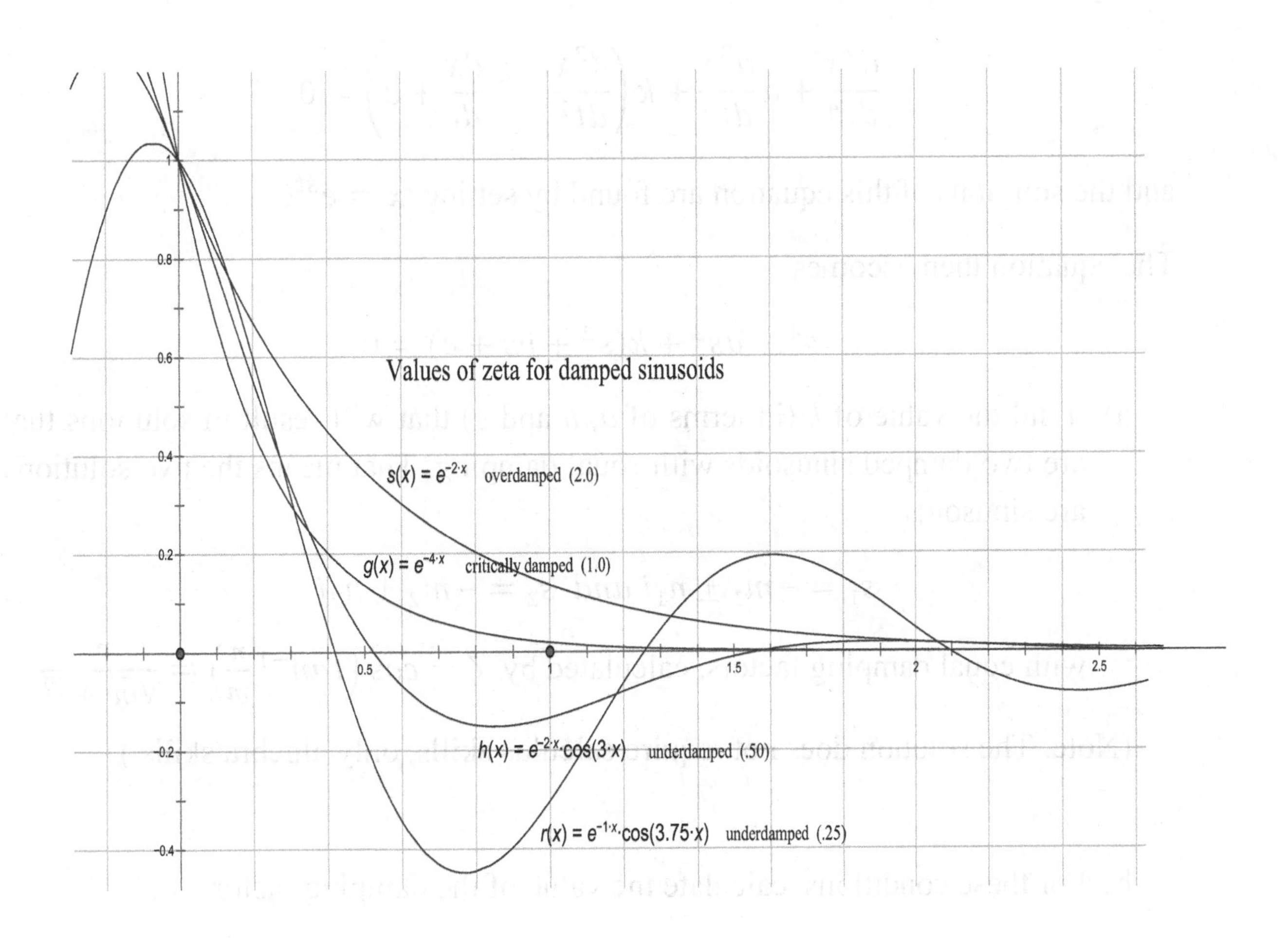

Informally, damping is a measure of how long it takes a sinusoid (sine wave or cosine wave) to reach and stay within a desired percentage of its initial or final value.

More formally, the damping factor (or damping ratio), symbolized by the Greek letter ζ, zeta, is an element of the solution

$$x(t) = e^{-\zeta\omega_0 t} cos\omega_0\sqrt{1-\zeta^2}t$$

to a second-order differential equation

$$\frac{d^2x}{dt^2} + 2\zeta\omega_0\frac{dx}{dt} + \omega_0^2(1-\zeta^2)x=0.$$

ω_0 is the undamped frequency (or the frequency when $\zeta = 0$).

Applying this to the foundational differential equation of automatic control system theory:

$$\frac{d^4x}{dt^4} + a\frac{d^3x}{dt^3} + k\left(\frac{d^2x}{dt^2} + b\frac{dx}{dt} + c\right) = 0$$

and the solutions of this equation are found by setting $x = e^{st}$.

The equation then becomes

$$s^4 + as^3 + k(s^2 + bs + c) = 0$$

a) Find the value of *k* (in terms of *a*, *b* and *c*) that will result in solutions that are two damped sinusoids with equal damping. That means the two solutions are sinusoids

$$s_1 = -m_1 \pm n_1 i \; and \; s_2 = -m_2 \pm n_2 i,$$

with equal damping factors, calculated by $\zeta = cos\left(tan^{-1}\frac{n}{m}\right) = \frac{m}{\sqrt{m^2+n^2}}$

(Note: The solution does not require calculus skills, only algebra skills.)

b) For these conditions, calculate the value of the damping factor, ζ.

For solutions with the same damping factor, the two sets of complex roots will be proportional and can be written as: $-m \pm ni$ and $r(-m \pm ni)$ where r is a constant, since the damping factor is

$$\zeta = cos\left(tan^{-1}\frac{n}{m}\right) = \frac{m}{\sqrt{m^2+n^2}}.$$

The equation can then be written as:

$$[(s+m+ni)(s+m-ni)]\big[\big(s+r(m+ni)\big)\big(s+r(m-ni)\big)\big] = 0$$

or

$$s^4 + 2m(r+1)s^3 + [(r^2+1)(m^2+n^2) + 4rm^2]s^2 +$$

$$2rm(m^2+n^2)(r+1)s + r^2(m^2+n^2)^2 = 0$$

Setting the coefficients of this equation with those of

$$s^4 + as^3 + k(s^2 + bs + c) = 0$$

we get

$$2m(r+1) = a$$

$$(r^2+1)(m^2+n^2) + 4rm^2 = k$$

$$2rm(m^2+n^2)(r+1) = kb$$

$$r^2(m^2+n^2)^2 = kc$$

Combining the first, third and fourth equations, yields the solution to part a)

$$k = \frac{ca^2}{b^2}.$$

Then, substituting into the second equation results in a quartic equation for r:

$$r^4 + (2 - \tfrac{a}{b})r^3 + (2 - \tfrac{2a}{b} + \tfrac{ab}{c})r^2 + (2 - \tfrac{a}{b})r + 1 = 0.$$

Dividing the equation by r^2, rearranging and setting $v = r + \frac{1}{r}$ we find (*LM* p 122)

$$v^2 - 2 + \left(2 - \tfrac{a}{b}\right)v + \left(2 - \tfrac{2a}{b} + \tfrac{ab}{c}\right) = 0.$$

Then $$v = \frac{\frac{a}{b} - 2 + \sqrt{\left(\frac{a}{b} + 2\right)^2 - \frac{4ab}{c}}}{2}$$ and

$$r = \frac{v + \sqrt{v^2 - 4}}{2}.$$

For the roots nearest the origin:

$$m = \frac{a}{2(r+1)} \text{ and } n = \sqrt{\frac{ac}{br} - m^2}$$

and the damping factor is

$$\zeta = \frac{m}{\sqrt{m^2+n^2}} = \frac{\sqrt{r}}{2(r+1)}\sqrt{\frac{ab}{c}}.$$

The solution to part b) is: $\zeta = \frac{\sqrt{r}}{2(r+1)}\sqrt{\frac{ab}{c}}$

where $$r = \frac{v + \sqrt{v^2 - 4}}{2}$$

and $$v = \frac{\frac{a}{b} - 2 + \sqrt{\left(\frac{a}{b} + 2\right)^2 - \frac{4ab}{c}}}{2}$$

Example $$\frac{d^4x}{dt^4} + 3\frac{d^3x}{dt^3} + k\left(\frac{d^2x}{dt^2} + \frac{dx}{dt} + 2.5\right) = 0$$

$$k = \frac{ca^2}{b^2} = \frac{2.5(3)^2}{(1)^2} = 22.5$$

$$.\ v = \frac{\frac{a}{b} - 2 + \sqrt{\left(\frac{a}{b} + 2\right)^2 - \frac{4ab}{c}}}{2} = \frac{\frac{3}{1} - 2 + \sqrt{\left(\frac{3}{1} + 2\right)^2 - \frac{4(3)(1)}{2.5}}}{2} = 2.7361$$

$$r = \frac{v + \sqrt{v^2 - 4}}{2} = \frac{2.7361 + \sqrt{2.7361^2 - 4}}{2} = 2.3016$$

$$m = \frac{a}{2(r+1)} = \frac{3}{2(2.3016 + 1)} = 0.4543$$

$$n = \sqrt{\frac{ac}{br} - m^2} = \sqrt{\frac{(3)(2.5)}{(1)(2.3016)} - (0.4543)^2} = 1.7471$$

$$\zeta = \frac{m}{\sqrt{m^2+n^2}} = \frac{.4543}{\sqrt{.4543^2+1.7471^2}} = .2514$$

or $$\zeta = \frac{\sqrt{r}}{2(r+1)}\sqrt{\frac{ab}{c}} = \frac{\sqrt{2.3016}}{2(2.3016+1)}\sqrt{\frac{(3)(1)}{2.5}} = .2517$$

Also $$\omega_0 = \sqrt{m^2 + n^2} = \sqrt{.4543^2 + 1.7471^2} = 1.7998$$

and $$\omega_1 = \omega_0\sqrt{1 - \zeta^2} = 1.7998\sqrt{1 - .2514^2} = 1.7420$$

The second set of roots is:

$$r(-m \pm ni) = 2.3016(-.4543 \pm 1.7471i) = -1.0475 \pm 4.033\, i$$

As expected, for these roots,

$$\zeta = \frac{m}{\sqrt{m^2+n^2}} = \frac{1.0475}{\sqrt{1.0475^2+4.0333^2}} = .2514$$

$$\omega_0 = \sqrt{m^2 + n^2} = \sqrt{1.0475^2 + 4.0333^2} = 4.1671$$

and $$\omega_1 = \omega_0\sqrt{1 - \zeta^2} = 4.1671\sqrt{1 - .2514^2} = 4.0333$$

When the two solutions are graphed on the same axes, we can see that the equal damping factors result in the curves' identical overshoots.

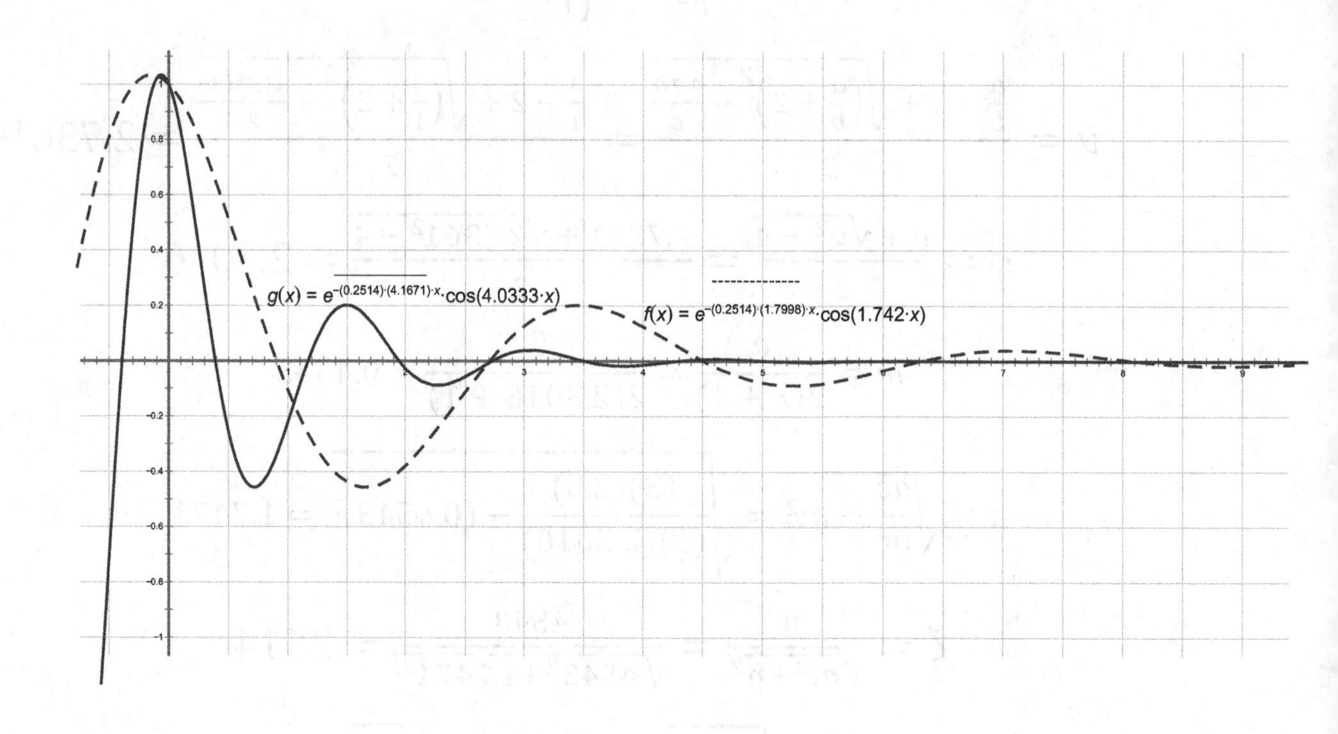

10 Raising a Complex Number to a Complex Number

Find the value of m and n if $(a+bi)^{(c+di)} = m + ni$

$(a+bi) = \sqrt{a^2+b^2}\; e^{itan^{-1}\frac{b}{a}}$ per Euler's formula and

$\sqrt{a^2+b^2} = e^{ln\sqrt{a^2+b^2}}$ by the definition of logarithms.

So, $(a+bi)^{(c+di)} = \left(e^{ln\sqrt{a^2+b^2}+itan^{-1}\frac{b}{a}}\right)^{(c+di)}$.

Multiplying the exponents gives

$$(a+bi)^{(c+di)} = e^{cln\sqrt{a^2+b^2}-dtan^{-1}\frac{b}{a}+i\left[ctan^{-1}\frac{b}{a}+dln\sqrt{a^2+b^2}\right]}$$

The final results are:

$$m = (a^2+b^2)^{\frac{c}{2}}\; e^{-dtan^{-1}\frac{b}{a}} cos\left(ctan^{-1}\frac{b}{a}+\frac{d}{2}ln(a^2+b^2)\right)$$

$$n = (a^2+b^2)^{\frac{c}{2}}\; e^{-dtan^{-1}\frac{b}{a}} sin\left(ctan^{-1}\frac{b}{a}+\frac{d}{2}ln(a^2+b^2)\right)$$

Example: $a=2$ $b=3$ $c=4$ $d=5$ so $(2+3i)^{(4+5i)} =$

$$(13)^2\; e^{-5tan^{-1}1.5}\, cos(4tan^{-1}1.5+2.5ln13))$$

$$+(13)^2\; e^{-5tan^{-1}1.5}\, sin(4tan^{-1}1.5+2.5ln13))i =$$

$$1.241\, cos(10.3435) + 1.241\, sin(10.3435)i = -.7530 - .9864i$$

10A Mandelbrot Fractals

The Mandelbrot Equation can be written:

$$z \rightleftharpoons z^2 + c \qquad \text{or} \qquad z_{n+1} = z_n^2 + c$$

where z and c are complex numbers.

1. What is a fractal?
2. Why are fractals important?
3. What is an iterative equation?
4. How does this equation work?
5. How does the imaginary term i affect this whole process?
6. What is the output of this equation?

10B Schrodinger (Wave) Equation

The time-dependent form of the Schrodinger Equation is:

$$i\hbar\frac{\partial}{\partial t}\Psi = \hat{H}\Psi$$

1. What does this equation represent or attempt to explain?
2. What do all the symbols represent?
3. In particular, what is the meaning of using the imaginary term i ?

10C Riemann's Hypothesis

Given that

$$\zeta(s) = \sum_{n=1}^{\infty}\frac{1}{n^s} = \frac{1}{1^s} + \frac{1}{2^s} + \frac{1}{3^s} + \ldots = 0$$

$$s = c + di = \frac{1}{2} + di$$

d has an infinite number of solutions

Prove that $c = ½$ for all solutions.

These three suggested projects likely represent the most interesting and important uses of imaginary numbers (in addition to electric circuit analysis and Maxwell's equations). Working on these projects will give students a far deeper appreciation of the potential for the use of imaginary numbers and how imaginary numbers can represent additional dimensions Imaginary numbers also play a role in modeling time delays in the real-time environment.

Mandelbrot Fractals incorporate both the mathematical and artistic realms. Videos of these fractals, with an increasing value of *n*, provide a surreal experience as we view what appear to be the inner workings of the universe.

Schrodinger's equation can help students understand electromagnetic wave propagation and gain an appreciation for multiple perpendicular axes using imaginary numbers.

Riemann's Hypothesis is the world's most famous unsolved problem. Computer algorithms have created billions of solutions to the equation without finding a single contradiction, yet a formal proof does not exist. Though fame and fortune await the person who solves it, there is concern that such a proof could lead to cracking encryption codes used for internet security.

Section III

Human Modeling Using Imaginary Numbers

Problem 11

11 Human Modeling Using Imaginary Numbers

We are accustomed to modeling tangible objects (like ourselves) in three physical and one time dimension. This enables us to locate ourselves in space, as well as determine our velocity and acceleration. But, there is more. How do we model our intangible, nonphysical parts? They are important, too.

Consider using imaginary numbers to form three mutually perpendicular axes. One axis could measure our logical thoughts, as in the left brain. One axis could measure our feelings, such as in our right brain. And one axis can be for our values and this axis could be vertical (such as having higher aspirations or lower morals, etc.). Imaginary time is the last axis and has been used by mathematicians such as Minkowski, working with relativity theory, and physicists such as Hawking, working on the big bang theory.

What equations would result from rotating one of the three intangible/imaginary axes such, as the x_i axis of values, toward the t_i imaginary time axis?

And, what would it mean, on the human level?

Advisory

For those readers who are not particularly facile in the concept of imaginary numbers, it is recommended that they read or reread this book's Forward.

Understanding how imaginary numbers developed and how their use has grown will help the reader understand and appreciate Section III, Problem 11.

Solution to Problem11

We start by comparing real- and imaginary time.

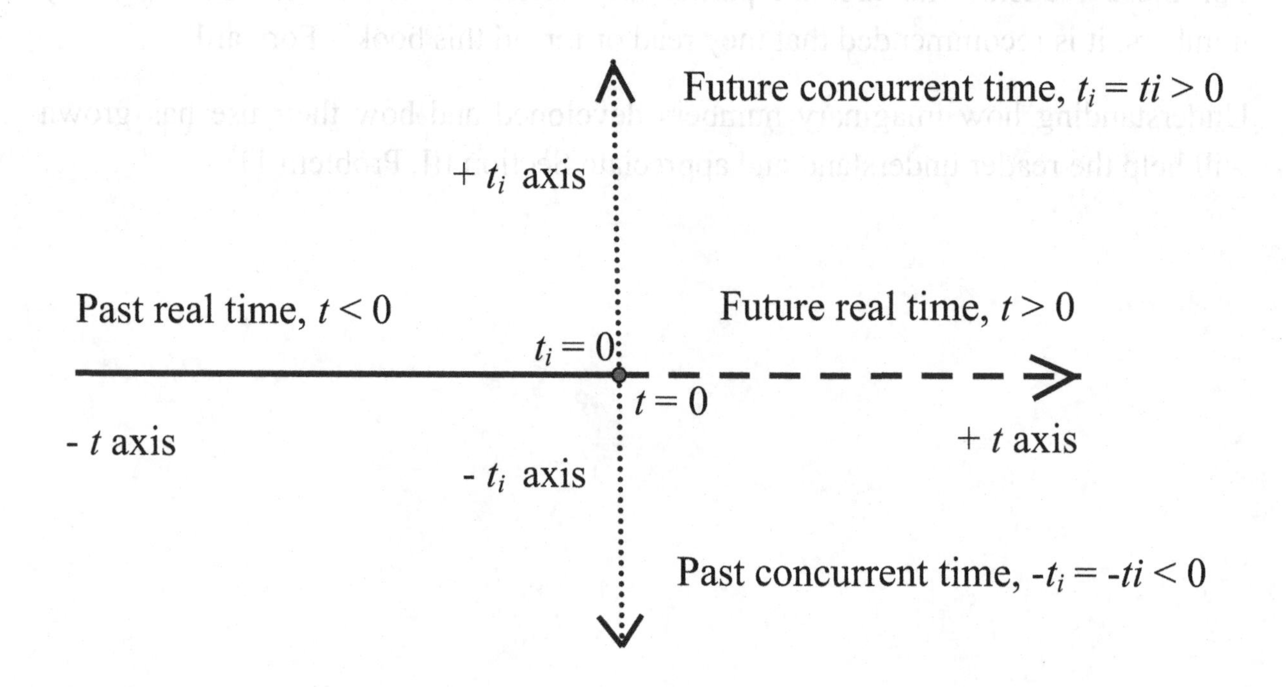

Real and Concurrent Time

Similarly, our three new imaginary axes are related to the real ones by

$x_i = ix$, $y_i = iy$ and $z_i = iz$. The result is a set of eight mutually perpendicular axes. So we can "locate" ourselves in an eight-term state vector. The good news is that this is easier to work mathematically than to visualize what it would physically look like.

An important concept we can address is "attitude." This word is a double entendre: on the one hand, it describes our approach or presentation to a situation or to people; on the other hand, it is the word used in mechanics to describe the angle between an axis and the direction an object is pointed. We can compare how the mechanics of people attitude compare with those of physical mechanics.

Let the Greek letter theta, θ, represent the real angle in the *xy* plane through which an object rotates (from the *x* axis toward the *y* axis). Using the standard rotation formula, the new axes' directions become

$$x' = x\,cos\theta - y\,sin\theta$$
$$y' = x\,sin\theta + y\,sin\theta \qquad \text{which are both real.}$$

Rotating our tangible axes means we will have a different view. In fact, all of our sensory inputs can change, and that can definitely have an effect on our values, ideas and feelings. But this is not necessarily the case. Changing our physical space may not have any consequential impact on our intangible components.

We can now calculate what happens when we rotate through the imaginary angle θ_i in the $x_i t_i$ plane. What this means is that we are reconsidering our values and priorities with an eye to the future, which concurrent imaginary time offers us. We use an imaginary angle since there is no physical rotation. This angle is of the kind that we speak of when we say, "turn our attention to" or "pivot our focus toward."

The equations begin comparably to the physical rotations, but then they develop remarkably differently.

$$x_i' = x_i\,cos\theta_i - t_i\,sin\theta_i = x_i\,cos(\theta i) - t_i\,sin(\theta i) = x_i\,cosh\,\theta - it_i\,sinh\,\theta$$
$$t_i' = x_i\,cos\theta_i + t_i\,sin\theta_i = x_i\,cos(\theta i) + t_i\,sin(\theta i) = x_i\,cosh\,\theta + it_i\,sinh\,\theta$$

Due to perpendicularity, $-it_i = t$, so the final results are

$$x_i' = x_i\,cosh\,\theta + t\,sinh\,\theta$$
$$t_i' = x_i\,cosh\,\theta - t\,sinh\,\theta$$

The rotation of the two imaginary axes is a combination of the imaginary axis, x_i, for values, and the real axis, *t*, for time.

We note that $\cos\theta, \sin\theta, \cosh\theta$ and $\sinh\theta$

are real numbers that are derived from trigonometric and hyperbolic formulas. They are scale factors and, at this time, are relatively unimportant.

Key Result

The key, significant result here is that x_i' is a combination of a portion of the original x_i and the impact of future real time t. And, future concurrent time t_i is a combination of a portion of the original x_i and the past impact of real time t. This shows that a change in our attitude about our values (and similarly our ideas and feelings) will certainly change us now and will provide us with motivation for greater growth in the future. Though changing our physical location may bring benefits, experience suggests that intangible attitude-changing results are of larger, longer lasting importance. Improvements through people attitude changes are indeed more impactful.

What more study can be done?

It is difficult to expand one's vision to a greater number of dimensions. See *Flatland*, by Abbott, which focused on imagining the third dimension above a two-dimensional plane And, as we have seen, Origami's use of the third dimension helps us solve geometry problems we could not do before.

Historically, the three spatial dimensions were considered to be the only dimensions until only recently. Since then, we have become comfortable with real time as a dimension, even though it only goes in one direction and we can't influence its speed (or even know what it is). With space and time interchangeable by relativity theory, imaginary number-based dimensions appear to be a natural consequence. We cannot see the three imaginary spatial dimensions of intangibles because they do not need any space. But as with *Flatland* and Origami, we allow ourselves to imagine how more dimensions can help us solve human questions from antiquity.

To date, physicists have proposed many theories of the universe with varying numbers of dimensions. And, their theories rely on having dimensions that are very small or curled up. One example has a curled-up dimension with a radius of 10^{-30} centimeters. So, this study's model with eight dimensions may yet become a subset of a model of the universe with even more dimensions. Perhaps, in time, physicists can test these theories in the laboratory.

Values, ideas and feelings can be better quantified both in polarity (positive or negative), as well as in size or impact. Such results could help inform social and legal codes to better align with better personal and communal life experiences.

Furthermore, humankind has debated whether a person's intangible qualities continue on after their death. On the one hand, they emanate from the brain, so maybe the intangibles die with the body. On the other hand, thoughts, feelings and values do have viability long after one's passing. Some people even claim to know about their previous lives. This analysis identifies the location of these intangibles in the universe, which adds to the affirmative plausibility argument of their survival.

Finally, I would add that future explorations into the realm of human intangibles open an exciting frontier that I believe will be very rewarding.